Hot and Cold Water Supply
An illustrated guide

Hot and Cold Water Supply
An illustrated guide

Prepared by

R.H. Garrett, EngTech, MIP RP
*Lecturer in Plumbing at Langley College of
Further Education
Past President of the National Association of
Plumbing Teachers
Plumber of the Year 1988*

for

British Standards Institution

OXFORD
BSP PROFESSIONAL BOOKS
LONDON EDINBURGH BOSTON
MELBOURNE PARIS BERLIN VIENNA

Copyright © British Standards Institution
1991

BSP Professional Books
A division of Blackwell Scientific
    Publications Ltd
Editorial offices:
Osney Mead, Oxford OX2 0EL
25 John Street, London WC1N 2BL
23 Ainslie Place, Edinburgh EH3 6AJ
3 Cambridge Center, Cambridge
    MA 02142, USA
54 University Street, Carlton
    Victoria 3053, Australia

First published 1991

Set by Best-set Typesetter Ltd
Printed and bound in Great Britain by
Cambridge University Press, Cambridge

DISTRIBUTORS

Marston Book Services Ltd
PO Box 87
Oxford OX2 0DT
(*Orders*: Tel: 0865 791155
         Fax: 0865 791927
         Telex: 837515)

USA
    Blackwell Scientific Publications, Inc.
    3 Cambridge Center
    Cambridge, MA 02142
    (*Orders*: Tel: (800) 759-6102)

Canada
    Oxford University Press
    70 Wynford Drive
    Don Mills
    Ontario M3C 1J9
    (*Orders*: Tel: (416) 441-2941)

Australia
    Blackwell Scientific Publications
    (Australia) Pty Ltd
    54 University Street
    Carlton, Victoria 3053
    (*Orders*: Tel: (03) 347-0300)

British Library
Cataloguing in Publication Data
Hot and cold water supply.
    1. Buildings. Private water supply
    systems. Installation, maintenance &
    repair
    I. Garrett, R. H.   II. British Standards
    Institution
    697.12

ISBN 0-632-02949-8

Contents

Acknowledgements

BSI wishes to thank Frank Young for his assistance and the following companies/organisations for their kind permission to use material.

Braithwaite Engineers Ltd
British Coal Corporation
British Gas Plc
Caradon Mira Ltd
Cistermiser Ltd
Conex-Sanbra Ltd
Construction Industry Training Board
Copper Development Association
Corgi
Danfoss Ltd
Department of the Environment
Ecowater Systems Ltd
F W Talbot & Co Ltd
George Fischer Sales Ltd
Gledhill Water Storage Ltd
Glow-Worm Ltd
Grandfoss Pumps
H Warner & Son Ltd
IMI Range Ltd
IMI Santon Ltd
IMI Yorkshire Fittings Ltd
Lancashire Fittings
Meter Options Ltd
North West Water Ltd
Polystel Ltd
Polytank Ltd
Reliance Water Controls
The Architectural Press Ltd
The Institute of Plumbing
Water Research Centre
Water Training International

Introduction

The development, construction, installation and maintenance of hot and cold water supply systems are vital areas of concern for public health. Water quality is by turns a political, environmental and technical issue. It is governed by legislation, water byelaws, building regulations and technical standards intended to safeguard quality.

This book is a thorough introduction and guide to hot and cold water supply. It is based on a British Standard, BS 6700, which is a specification for design, installation, testing and maintenance of water supply services for domestic buildings. It also includes information on the law and on the water byelaws. It is an invaluable work of reference for designers, installers and contractors who work on domestic water supply. It is essential reading for students of plumbing – especially those who are working for City and Guilds or BTech examinations. It is also an important aid to technical staff of water undertakers, structural surveyors, architects, building contractors and property services managers.

At present it is not a legal requirement that anyone installing or repairing domestic water services should be properly qualified. However, the installer or repairer can be prosecuted for offences under the water byelaws or could be prosecuted under civil law by a householder. With increasing pressure to maintain the highest standards of health and safety and to prevent waste, misuse or contamination of water it is important that technical knowledge is up-to-date.

There is also a strong movement to make water supply practice consistent throughout Europe. In continental Europe hot water systems are supplied from mains pressure rather than from storage. This book helps practitioners to come to terms with these changes.

Hot and Cold Water Supply was written for the British Standards Institution by Bob Garrett, and has benefited from the assistance and comments of many other plumbers, designers and teachers. A principle of *Hot and Cold Water Supply* is to support technical information with illustrations and examples. This is especially useful for students and for those who are engaged in training. It is an immensely practical book which makes easy the job of turning theory into action.

Derek Prior
Head of Education, BSI

Chapter 1
General considerations

1.1 Legislation

In England and Wales, water supply is governed largely by water byelaws, made by water undertakers under section 17 of the Water Act 1945, for the prevention of waste, undue consumption, misuse or contamination of water supplied by the undertaker. These byelaws are expected to be replaced by regulations made by the Secretary of State for the Environment in due course. They may also have to be amended as a result of the Water Act 1989, and as a consequence of future developments in the EEC.

Under the 1945 Act the following terms are used.

'waste' means water which flows away unused, e.g. from a dripping tap or hosepipe left running when not in use, or from a leaking pipe.

'undue consumption' means water used in excess of what is needed, e.g. full bore tap to wash hands when half or quarter flow will suffice, or automatic flushing cistern that flows even when urinals are not in use (at night).

'misuse' means water used for purposes other than that for which it is supplied, e.g. use of garden sprinkler when paying only for domestic use or taking supply from domestic premises for industrial or agricultural use.

'contamination' means pollution of water by any means, e.g. by cross connection between public and private supplies or by backflow through backsiphonage.

'erroneous measurement' means incorrect meter reading, e.g. connections made which may not be detected by the meter.

Scotland has byelaws made under section 70 of the Water (Scotland) Act 1980 that are virtually identical with the byelaws which apply in England and Wales.

Water supply in Northern Ireland has been controlled by Water Regulations made under Article 40 of the Water and Sewerage Services (Northern Ireland) Order 1973. New regulations (on a par with byelaws in Britain) are expected shortly.

The new water byelaws were finally approved and came into force simultaneously throughout Britain on 1 January 1989.

A building owner or occupier can demand a supply of water for domestic purposes provided he has complied with the relevant requirements of the Water Act 1989 and the installation satisfies the requirements of the water byelaws.

Whilst it is the duty of the water supplier to accede to the owner's demand, he must also uphold the requirements of the water byelaws and has the right to refuse connection to the mains of any new installation which is not in compliance with them.

To avoid unnecessary dispute with water undertakers, and perhaps lengthy legal proceedings, it is advisable to consult water undertakers about their byelaws at an early stage, and particularly their requirements arising from local water or soil characteristics.

Although there is no legal requirement for a person installing or repairing water services to be suitably qualified, anyone who carries out work of this kind can be prosecuted for an offence against current water byelaws. It should be noted that householders also commit an offence under the Act if they 'use' a fitting which does not comply with the byelaws. Knowledge of the Act is therefore desirable for both installer and user.

Particular note should be taken of byelaws 97 and 98 which require five days' notice, in writing, before commencing work on any of the following:

installation of or alteration to a:

- bidet;
- flushing cistern;
- hose union tap or tap to which a hose can be connected;
- water fitting which could cause contamination through backflow;

or before:

- backfilling any excavation in which a pipe is laid;
- threading a pipe through a duct into a building;
- embedding a pipe in a solid wall or floor;
- laying a pipe underground by mole plough or similar apparatus.

(In Scotland five days' notice must be given before installing or altering any water fitting.)

Byelaw 99 prescribes that any person (installer or user) who contravenes water byelaws can be fined, if convicted, up to:

(1) £400 for each offence (in Scotland level four on the standard scale);
(2) £40 for each day during which each offence continues after conviction (in Scotland £50 per day).

It should be added that water byelaws are *not* made for the specific protection of people or property, but solely for the prevention of waste, undue consumption, misuse and contamination of water supplied by water undertakers.

Health and safety in respect of water systems is covered by the Building Regulations. These include:

- the provision of baths, showers and other appliances;
- the provision of hot and cold water;

- precautions against explosion in hot water systems, including unvented hot water systems;
- controls on certain space and water heating installations;
- insulation of hot water heating and supply pipes and hot store vessels;
- provision for effective cleansing of water closets and urinals.

Guidance on particular water byelaw matters may be sought from local water undertakers. Their inspectors are trained in byelaw application.

The Water Research Centre Water Byelaws Advisory Service has published a guide (*Water Supply Byelaws Guide*, second edition (1989) published by the Water Research Centre in association with Ellis Horwood) to the application and interpretation of water byelaws. It contains useful information and background knowledge for those concerned with water services.

1.2   Scope of the standard

BS 6700 specifies requirements and gives recommendations for the design, installation, alteration, commissioning and testing, and maintenance of hot water supply, cold water supply and frost precautions.

The following are not included although, in parts, the standard may apply to them: hot water systems whose temperature exceeds 100°C; central heating systems; fire fighting; and water for industrial purposes.

1.3   Testing and approval of water fittings

The Water Research Centre also operates a voluntary national scheme for the testing and approval of water fittings. Fittings which pass the Centre's tests are listed in its publication *Water Fittings and Materials Directory*, together with the names and addresses of manufacturers and any applicable installation requirements. Thus, the connection and use of any listed fitting carries with it certainty of acceptance by water undertakers.

Many of the changes incorporated in the current water byelaws in Britain arose from the *Backsiphonage Report* of 1974 and from requests from various sources for Britain to permit the supply of hot water directly under mains pressure. These changes are reflected in this book.

The main changes incorporated in the new water byelaws include:

- Stricter requirements for cold water storage cisterns to maintain water quality. Cisterns for drinking water and water used for cooking must be treated as 'protected' cisterns. These are required to have securely fixed and close-fitting lids, and be sealed to exclude light, vermin and insects. Vents and overflows to the cistern must be screened.

  'Portsmouth' type floatvalves to BS 1212: Part 1 are prohibited unless used with a double check valve arrangement.
- Unvented hot water storage heaters over 15 litres are now permitted (provided the appropriate backflow devices are fitted).

- Connection is now permitted, in ways which were previously illegal, of bidets with ascending sprays.
- WCs to have a maximum flush of 7.5 l instead of 9.5 l. A timetable has been drawn up to permit compliance by January 1993. Also, reduced flushing arrangements now apply to urinals.
- Lead pipes are not permitted for hot and cold water supplies, and lead-free solder must be used for all parts of a system from which water may be consumed.
- New requirements have been introduced for the prevention of backflow and backsiphonage and the avoidance of cross-connections.

Guidance notes from the Water Research Centre based on guidance from the Department of the Environment draw attention to:

'the advantages of specifying or employing installers and operatives who are enrolled on the Register of Plumbers monitored and supervised by the Institute of Plumbing; those entered on the register are identified by the title "Registered Plumber" and the letters "RP" after their names'.

These notes further advise that where work is carried out by '...a Registered Plumber...the consumer shall not be considered as having contravened any requirement of byelaws 2 or 3(a)'.

1.4 Definitions

From BS 6700

The following definitions are used:

building   any permanent, temporary, movable or immovable structure (including a floating structure), connected to the water supplier's mains.

cavity wall   any wall, whether structural or partition, which is formed by two upright parts of similar or dissimilar building materials suitably tied together with a gap formed between them even though this may be filled with insulating material.

chase   a recess, cut or preformed so that it does not affect the strength of the structure in which it is formed, and has a minimum effect on any thermal or acoustic insulation. A chase is designed to accommodate water pipes and fittings which can be accessible by removal of a cover or covers.

cover   a panel or sheet of rigid material fixed over a chase, duct or access point, strong enough to withstand surface loadings appropriate to its position. A cover may be plastered or screeded over except where it provides access to joints or changes of direction (e.g. at an inspection access point).

duct   an enclosure designed to accommodate water pipes and fittings and other services, if required, constructed so that one may gain access to the interior either throughout its length or at specified points, by removal of a cover or covers.

dwelling   premises providing living accommodation including: any house, flat, maisonette, caravan, vessel, boat or houseboat connected to the water supplier's mains.

inspection access point   a position of access to a duct or chase which allows for a pipe or pipes to be inspected by removing a cover which is fixed by removable fastenings but does not necessitate the removal of surface plaster, screed or continuous surface decoration.

removable fastenings   fastenings which can be removed readily and replaced without causing damage. These may include turn buckles, clips, magnetic or touch latches, coin operated screws and conventional screws, but exclude nails, pins or adhesives.

sleeve   an enclosure of tubular or other section of suitable material designed to provide a space through an obstruction to accommodate a single water pipe, e.g. a pipe passing through a wall.

tap size designations   numbers directly related to the nominal size of the thread on the inlet of the tap. This nominal size is the same as the nominal size in inches before metrication, e.g. $\frac{1}{2}$ nominal size tap means a tap with an inlet having a thread to BS 21.

walkway or crawlway   an enclosure similar to a duct, but large enough to allow a person access to the interior through doors or manholes and which will accommodate water pipes and fittings and other services if required.

water supplier   any water supply organisation, including Water Authorities, Regional and Islands Councils, and Water Companies. Also referred to as a water undertaker.

From the Model Water Byelaws

A further list of definitions, from the Model Water Byelaws, are as follows:

backflow   flow in a direction contrary to the intended normal direction of flow.

backsiphonage   backflow caused by the siphonage of liquid from a cistern or appliance into the pipe feeding it.

boiler   an enclosed vessel in which water is heated by the direct application of heat.

cistern   a fixed container for holding water at atmospheric pressure.

closed circuit   any system of pipes and other water fittings through which water circulates but from which no water is drawn for use, and includes any vent pipe fitted thereto but not the feed cistern or the cold feed pipe.

combined feed and expansion cistern   a cistern for supplying cold water to a hot water system without a separate expansion cistern.

communication pipe   that part of a service pipe which is vested in the undertakers.

cylinder   a cylindrical closed vessel capable of containing water under pressure greater than atmospheric pressure.

distributing pipe   any pipe (other than an overflow pipe or a flush pipe) conveying water from a storage cistern, or from a hot water apparatus supplied from a feed cistern, and under pressure from that cistern.

double feed indirect cylinder   an indirect cylinder which has separate cold feed pipe connections for both the primary circuit and the secondary circuit.

expansion cistern   a cistern connected to a water heating system which accommodates the increase in volume of water in that system when it is heated from cold.

feed cistern   any storage cistern used for supplying cold water to a hot water apparatus, cylinder or tank.

float-operated valve   a valve for controlling the flow of water into a cistern, the valve being operated by the vertical movement of a float riding on the surface of the water.

flushing cistern   a cistern provided with a device for discharging the stored water rapidly into a water-closet pan or urinal.

flush pipe   a pipe for conveying water from a flushing cistern to a water-closet pan or urinal.

flushing trough   a flushing apparatus which combines several discharging units in one long cistern body to allow more frequent flushing of two or more water-closet pans.

indirect cylinder   a hot water cylinder in which the stored water is heated by a primary heater through which hot water is circulated from a boiler or gas circulator without mixing of the primary and secondary water taking place.

instantaneous water heater   an appliance in which water is immediately heated as it passes through the appliance.

primary circuit   an assembly of pipes and fittings in which water circulates between a boiler or other water heater and the primary heater inside a hot water storage vessel.

primary heater   a heater mounted inside a hot water storage vessel for the transfer of heat to the stored water from circulating hot water.

secondary circuit   an assembly of pipes and fittings in which water circulates in distributing pipes to and from a water storage vessel.

secondary system   that part of a hot water system comprising the cold feed pipe, any storage cistern, water heater and flow and return pipework from which hot water for use is conveyed to all points of draw-off.

service pipe   so much of any pipe for supplying water from a main to any premises as is subject to water pressure from that main, or would be so subject but for the closing of some valve.

servicing valve   a valve for shutting off the flow of water in a pipe connected to a water fitting to facilitate the maintenance or servicing of that fitting.

single feed indirect cylinder   an indirect cylinder which has only one cold feed pipe connection to supply both the primary and secondary waters, so designed that the formation of an air seal during filling prevents mixing and accommodates expansion of the primary water.

spill-over level   the level at which the water in a cistern or vessel will first spill over if the inflow exceeds the outflow through any outlet and any overflow pipe.

stopvalve   a valve, other than a servicing valve, fitted in a pipeline for controlling or stopping at will the flow of water.

storage cistern   any cistern storing water for subsequent use, other than a flushing cistern.

supply pipe   so much of any service pipe as is not a communication pipe.

tank   a non-cylindrical closed vessel capable of containing water under pressure greater than atmospheric pressure.

unvented primary circuit   a primary circuit which is not provided with a vent pipe.

vent pipe   a pipe open to the atmosphere and used in connection with a hot water system for the escape of air or steam.

vented primary circuit   a primary circuit which is provided with a vent pipe.

warning pipe   an overflow pipe so fixed that its outlet, whether inside or outside the building, is in a conspicuous position where the discharge of water can be readily seen.

washing trough   a wash basin, wash trough or sink measuring internally more than 1.2 m over its longest or widest part, at which two or more persons can wash at the same time.

water supplied for domestic purposes   water supplied by the undertakers for drinking, washing, cooking and sanitary purposes and includes, where water is drawn from a tap inside a dwelling and no hosepipe or similar apparatus is used, watering a garden and washing vehicles kept for private use.

water supplied for non-domestic purposes   water supplied by the undertakers for domestic purposes which has been drawn off for use, water which is unfit for human consumption and water from any source other than the undertakers' mains.

water fittings   includes pipes (other than mains), taps, cocks, valves, ferrules, meters, cisterns, baths, water-closets, soilpans and other similar apparatus used in connection with the supply and use of water.

Graphical symbols

See figure 1.1.

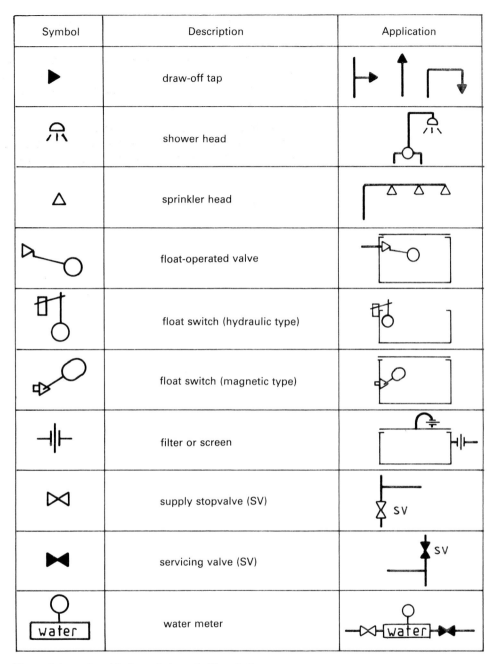

Symbol	Description	Application
▶	draw-off tap	
	shower head	
△	sprinkler head	
	float-operated valve	
	float switch (hydraulic type)	
	float switch (magnetic type)	
‖	filter or screen	
⋈	supply stopvalve (SV)	SV
⋈	servicing valve (SV)	SV
water	water meter	water

Figure 1.1   Graphical symbols and abbreviations

continued

Symbol	Description	Application
⋈	draining valve (BS 1192) (drain valve) (drain cock)	
⊏	draining valve (abbreviated version used in this book)	
⊥	line strainer	
⋈	pressure reducing valve (small end denotes high pressure)	
⊖	expansion vessel	
⋈	pressure relief valve	
▷	check valve or non-return valve (NRV)	
▷⊤▷	double check valve assembly	
⋈	combined check and anti-vacuum valve (check valve and vacuum breaker)	
△	air inlet valve	

Figure 1.1   Graphical symbols and abbreviations continued

continued

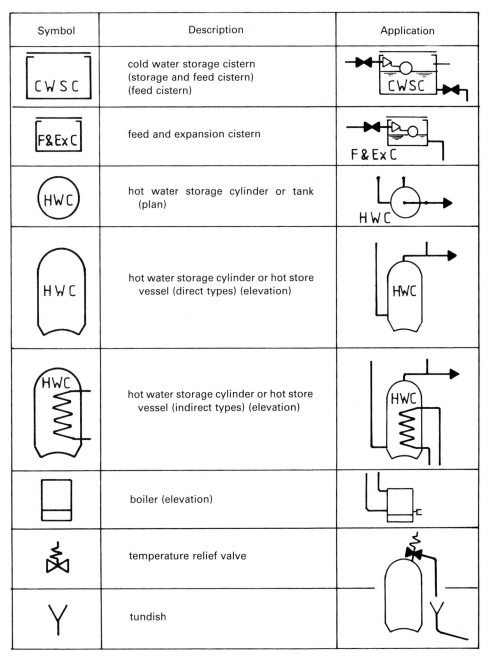

Symbol	Description	Application
C W S C	cold water storage cistern (storage and feed cistern) (feed cistern)	
F&ExC	feed and expansion cistern	
HWC	hot water storage cylinder or tank (plan)	
HWC	hot water storage cylinder or hot store vessel (direct types) (elevation)	
HWC	hot water storage cylinder or hot store vessel (indirect types) (elevation)	
	boiler (elevation)	
	temperature relief valve	
Y	tundish	

Figure 1.1   Graphical symbols and abbreviations
continued

continued

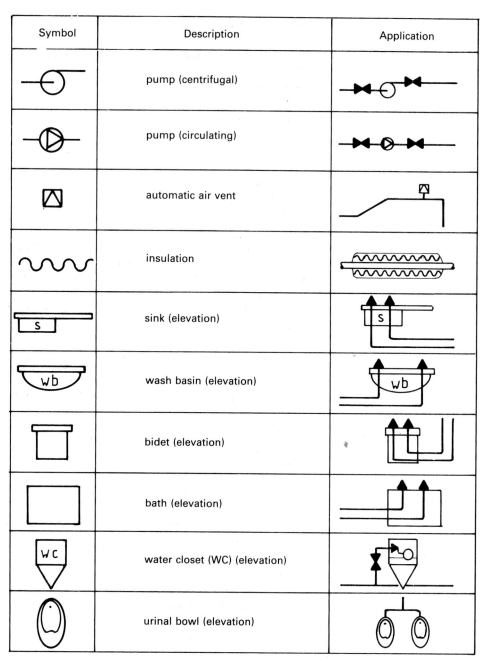

Symbol	Description	Application
	pump (centrifugal)	
	pump (circulating)	
	automatic air vent	
	insulation	
s	sink (elevation)	
wb	wash basin (elevation)	
	bidet (elevation)	
	bath (elevation)	
WC	water closet (WC) (elevation)	
	urinal bowl (elevation)	

**Figure 1.1   Graphical symbols and abbreviations
continued**

1.5 Materials

Selection and use of materials is dealt with in greater depth in chapter 11.

Choice of material for a particular water installation is determined by the following factors:

- effect on water quality;
- cost, service life and maintenance needs;
- for metallic pipes internal and external corrosion, (particularly from certain waters);
- compatibility of different materials;
- ageing, fatigue and temperature effects, especially in plastics;
- mechanical properties and durability.

Pipes and fittings should be used only within the limits stated in relevant British Standards and in accordance with any manufacturers' recommendations. Installations should be capable of operating effectively under the conditions they will experience in service.

Pipes, joints and fittings should be capable of withstanding sustained temperatures as shown in table 1.1, without damage or deterioration.

Table 1.1   Temperature limits for pipes and fittings

Cold water installations	40°C
Hot water installations	70°C (with occasional short term excursions up to 100°C)
Draw-off taps	65°C
Discharge pipes connected to temperature or expansion relief valves in unvented hot water systems	125°C

Pipes, joints and fittings of dissimilar metals should not be connected together unless precautions are taken to prevent corrosion. This is particularly important in below ground installations (see table 1.2).

1.6 Initial procedures

Preliminary design factors are as follows:

- flow requirements;
- location of available water supply;
- quality, quantity and pressure of available supply;
- quantity of storage required;
- requirements of building regulations and water byelaws;
- ground conditions;
- liaison with other parties.

Pipe sizes

Since about 1970 pipe sizes have gradually been changing from imperial to metric measurements. Pipes in some materials were metricated quickly, e.g. copper, while others even now retain their imperial identification.

With metrication came a move away from designating pipe sizes by the inside diameter towards the use of the outside diameter.

The result of these changes is a complicated system in which some pipes are designated by their inside diameter and others by their outside diameters, and some have been metricated and others have not.

Table 1.2 provides a comparison between the sizes of pipes of different materials, using inside diameters as a base, because it is the inside diameter which determines the water carrying capacity of the pipe.

Table 1.2   Equivalent pipe sizes   **Rigid pipes**

Comparative interval size in millimetres	Copper* to BS 2871 (table X)		Stainless Steel to BS 4127: Part 2		Steel (screwed) (galvanized or black)* to BS 1387 (medium grade)			Grey iron to BS 4622		Ductile iron to BS 4772		Asbestos cement* to BS 484 (Class 25)	
	BS nominal size		BS nominal size		BS nominal size		Thread designation	BS nominal diameter		BS nominal diameter		BS nominal diameter	
	ID mm	OD mm	ID mm	OD mm	ID mm	OD mm	BS21 inches	ID mm	OD mm	ID mm	OD mm	ID mm	OD mm
4.5	4.8	6	4.8	6	—	—	—	—	—	—	—	—	—
6	6.8	8	6.8	8	—	—	—	—	—	—	—	—	—
8	8.8	10	8.8	10	8	13.6	$\frac{1}{4}$	—	—	—	—	—	—
10	10.8	12	10.8	12	10	17.1	$\frac{3}{8}$	—	—	—	—	—	—
13	13.6	15	13.6	15	15	21.4	$\frac{1}{2}$	—	—	—	—	—	—
15	16.4	18	16.6	18	—	—	—	—	—	—	—	—	—
20	20.2	22	20.6	22	20	26.9	$\frac{3}{4}$	—	—	—	—	—	—
20	26.2	28	26.4	28	25	33.8	1	—	—	—	—	—	—
32	32.6	35	33	35	32	42.5	$1\frac{1}{4}$	—	—	—	—	—	—
40	39.6	42	39.8	42	40	48.4	$1\frac{1}{2}$	—	—	—	—	—	—
50	51.8	54	—	—	50	59.3	2	—	—	—	—	50	69
63	64.6	67	—	—	65	80.1	$2\frac{1}{2}$	—	—	—	—	—	—
75	73.1	76.1	—	—	80	88.8	3	80	98	80	98	75	96
100	105	108	—	—	100	113.9	4	100	118	100	118	100	122
125	130	133	—	—	125	139.6	5	—	—	—	—	125	—
150	155	159	—	—	150	165.1	6	150	170	150	170	150	177
200	—	—	—	—	—	—	—	200	222	200	222	200	240
250	—	—	—	—	—	—	—	250	274	250	274	250	295
300	—	—	—	—	—	—	—	300	326	300	326	300	356
								†	†	†	†	†	†

Note   Some intermediate sizes have been omitted
* In some materials only one grade is shown.
† Larger sizes than 300 mm have been excluded

Flexible pipes

uPVC* to BS 3505 (Class E) ID inches	OD mm	ID mm	Propylene copolymer to BS 4991 ID inches	OD mm	Polyethylene to BS 6572 medium density (BLUE) ID mm	BS nominal size OD mm	to BS 6730 medium density (BLACK) ID mm	BS nominal size OD mm	to BS 1972 low density Class C OD mm	to BS 1384 high density BS nominal size ID inches	Class C OD mm	Comparative interval size in inches
—	—	—	—	—	—	—	—	—	—	—	—	$\frac{1}{8}$
—	—	—	—	—	—	—	—	—	—	—	—	$\frac{1}{4}$
—	—	—	$\frac{1}{4}$	13.6	—	—	—	—	—	—	—	$\frac{3}{8}$
—	—	—	$\frac{3}{8}$	17.1	—	—	—	—	17.1	$\frac{3}{8}$	17.1	$\frac{1}{2}$
$\frac{3}{8}$	17.1	15.3	$\frac{1}{2}$	21.3	15.1	20	15.1	20	21.3	$\frac{1}{2}$	21.3	$\frac{3}{4}$
$\frac{1}{2}$	21.3	17.5	$\frac{3}{4}$	26.7	20.4	25	20.4	25	26.7	$\frac{3}{4}$	26.7	1
$\frac{3}{4}$	26.7	21.7	1	33.5	26	32	26	32	33.5	1	33.5	$1\frac{1}{4}$
1	33.5	28.3	$1\frac{1}{4}$	42.2	—	—	—	—	42.3	$1\frac{1}{4}$	42.3	$1\frac{1}{2}$
$1\frac{1}{4}$	42.2	34.8	$1\frac{1}{2}$	48.3	40.8	50	40.8	50	48.3	$1\frac{1}{2}$	48.3	2
$1\frac{1}{2}$	48.2	41.0	2	60.3	51.4	63	51.4	63	60.3	2	60.3	$2\frac{1}{2}$
2	60.3	51.3	$2\frac{1}{2}$	75.3	—	—	—	—	—	—	—	3
—	—	—	3	88.9	—	—	—	—	—	—	—	4
3	88.9	76.5	4	114.3	—	—	—	—	—	—	—	5
4	114.3	97.7	—	—	—	—	—	—	—	—	—	6
5	140.2	120	6	168.3	—	—	—	—	—	—	—	8
6	168.2	144	8	219.5	—	—	—	—	—	—	—	10
8	219.1	190.9	10	273.4	—	—	—	—	—	—	—	12
10	273	138	12	323.8	—	—	—	—	—	—	—	
12	323.8	282.2	†	†	—	—	—	—	—	—	—	
†	†	†										

Note Some intermediate sizes have been omitted
* In some materials only one grade is shown.
† Larger sizes than 300 mm have been excluded

Availability of water supplies

Water supplies are available from:

(1) Nearby public water main at cost to owner or within contract price. If not readily available, the water undertaker must provide mains at the expense of the owner or applicant.
Note Mains may not be laid until road line and kerb level are permanently established.
(2) Suitable and available supply pipe (not favoured by water suppliers).
(3) Private source. Consider the condition and purity of the water. Chemical and bacterial analyses are advisable.
Note If private supply and public supply are taken to a property, the water undertaker must be informed and byelaws complied with.

Water suppliers will need full details of non-domestic water supply requirements to assess the likely demand and effects on water mains and other users in the locality (see figure 1.2).

Ground conditions to be considered

(1) Likelihood of contamination of site (local authorities may provide information).
(2) Likelihood of subsidence and other soil movement from:
    (i) mining;
    (ii) vibration from traffic (consider increased depth);
    (iii) moisture swelling and contraction;
    (iv) building settlement;
(3) When laying underground service pipes make allowances for pipe materials, jointing and methods of laying. Consider methods of passing pipes through walls. Pipes must be free to deflect.

As far as possible underground pipes:

• should *NOT* be laid under surfaced footpaths or drives;
• should be laid at right angles to the main;
• should be laid in straight lines to facilitate location for repairs, but with slight deviation to adjust to minor ground movement.

Liaison and consultation

From an early stage in the design process, the designer should consult with others involved in the design, installation and use of the system (see figure 1.3).

The designer or installer should provide full working drawings, including precise location of all pipe runs, method of ducting, description of all appliances, valves and other fittings, methods of fixing, protections and precautions.

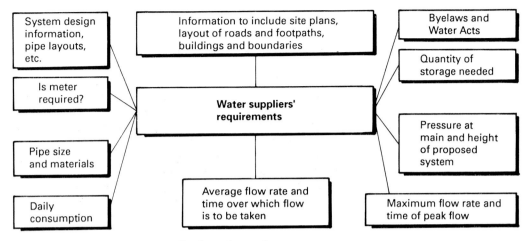

Figure 1.2   Water suppliers' requirements

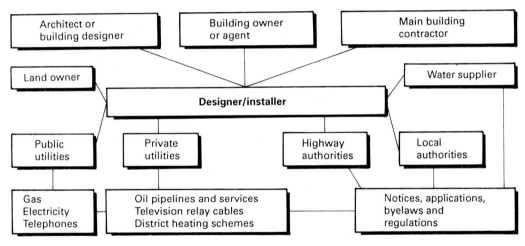

Figure 1.3   Designers'/installers' liaisons

The programme of work should consider:

- method of construction;
- sequence of events including handover to owner;
- coordination of services;
- time needed for construction and services works.

Chapter 2
Cold water supply

2.1 Drinking water

Under the Public Health Acts, every dwelling is required to have a potable (drinkable) water supply, and the most important place to provide drinking water in dwellings is at the kitchen sink (see figure 2.1). However, because there is a likelihood that all taps in dwellings will be used for drinking, they should all be connected in such a way that the water remains in potable condition. This means that all draw-off taps in dwellings should either be connected direct from the mains supply, or from a storage cistern that is 'protected'.

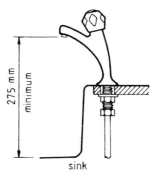

275 mm minimum

Positioned so that outlet is at least 275 mm above bottom of bowl, to allow buckets and other utensils to be filled.

Outlets designed to make hosepipe connections difficult.

sink

Figure 2.1   Tap at kitchen sink

Drinking water supplies should also be provided in suitable and convenient locations in offices and other buildings, particularly where food and drink is prepared or eaten. If no such locations exist, drinking water should be provided near but preferably not in toilets. However, drinking water fountains may be installed in toilet areas, provided they are sited well away from WCs and urinals and comply with the requirements of BS 6465: Part 1.

To avoid stagnation drinking water points should not be installed at the ends of long pipes where only small volumes of water are likely to be drawn off.

As far as possible pipe runs to drinking water taps should not follow the routes of space heating or hot water pipes or pass through heated areas. Where this is unavoidable, both hot and cold pipes should be insulated.

2.2 Types of system

Systems in dwellings

Water may be supplied to cold taps either directly from the mains via the supply pipe or indirectly from a protected cold water storage cistern. In some cases a combination of both methods of supply may be the best arrangement.

Factors to consider when designing a cold water supply system are shown in figures 2.2 and 2.3.

Cistern omitted where hot water is supplied from: (a) an unvented hot water system; or (b) mains-fed instantaneous water heater.

High pressure supply is more suitable for instantaneous type shower heaters, hose taps and mixer fittings used in conjunction with a high pressure (unvented) hot water supply.

Expensive dual-flow mixer fittings required if used in conjunction with a low pressure (vented) hot water supply.

All taps are supplied under mains pressure and are therefore suitable for drinking and food preparation.

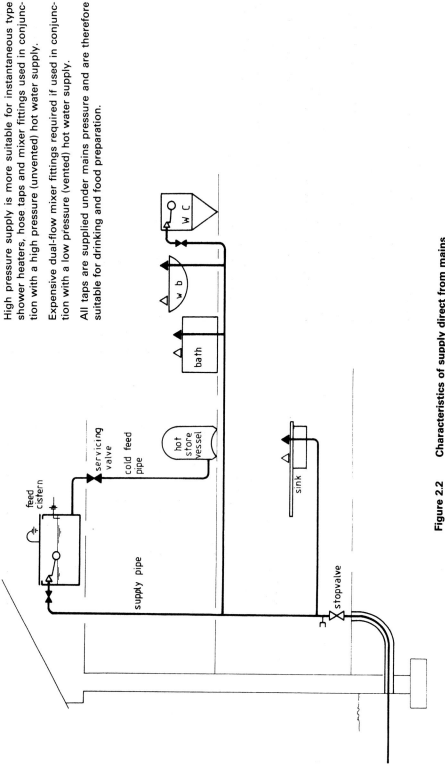

Figure 2.2   Characteristics of supply direct from mains

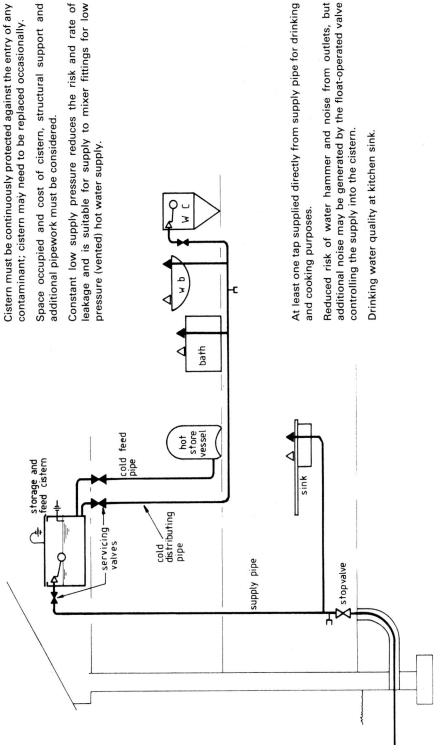

Supply pipe must be protected against backflow from cistern.

Risk of frost damage in roof space.

Reserve supply of water available in case of mains failure.

Pressure available from storage may not be sufficient for some types of tap or shower.

Cistern must be continuously protected against the entry of any contaminant; cistern may need to be replaced occasionally.

Space occupied and cost of cistern, structural support and additional pipework must be considered.

Constant low supply pressure reduces the risk and rate of leakage and is suitable for supply to mixer fittings for low pressure (vented) hot water supply.

At least one tap supplied directly from supply pipe for drinking and cooking purposes.

Reduced risk of water hammer and noise from outlets, but additional noise may be generated by the float-operated valve controlling the supply into the cistern.

Drinking water quality at kitchen sink.

storage and feed cistern

cold feed pipe

hot store vessel

servicing valves

cold distributing pipe

sink

supply pipe

stopvalve

w c

w b

bath

Figure 2.3   Characteristics of supply via storage cistern

Systems in buildings other than dwellings

In the case of small buildings where the water consumption is likely to be similar to that of a dwelling, the characteristics in figures 2.2 and 2.3 should be considered.

For larger buildings such as office blocks, hostels and factories it will usually be preferable for all water, except drinking water, to be supplied indirectly from a cold water storage cistern or cisterns. Drinking water should be taken directly from the water supplier's main wherever practicable (see figure 2.4).

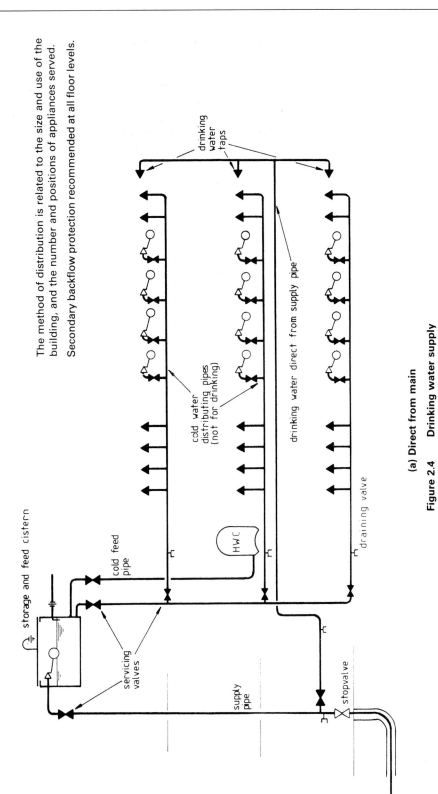

The method of distribution is related to the size and use of the building, and the number and positions of appliances served.

Secondary backflow protection recommended at all floor levels.

(a) Direct from main

Figure 2.4   Drinking water supply

continued

For use where a separate drinking water supply cannot be provided.

Secondary backflow protection recommended at all floor levels.

storage and feed cistern 'protected' type

cold feed pipe

servicing valves

supply pipe

stopvalve

cold water distributing pipes

all taps are suitable for drinking purposes

draining valve

HWC

(b) Alternative system from storage

Figure 2.4   Drinking water supply
continued

Pumped systems

Where the height of the building lies above statutory levels or when the available pressure is insufficient to supply the whole of a building and the water supplier is unable to increase the supply pressure in the supplier's mains, consideration should be given to the provision of a pumped cold water supply cistern.

2.3 Storage cisterns

Storage cisterns and lids for domestic purposes should not impart taste, colour, odour or toxicity to the water, nor promote microbial growth.

Cistern requirements

The following cistern requirements are noted (see figure 2.5):

- Materials should be suitable for maintaining potable water quality and must not deform unduly in use.
- All cisterns and pipes should be insulated against the effects of frost or heat.
- Cisterns should be situated away from heat.
- Access should be provided for inspection and maintenance (both internally and externally).

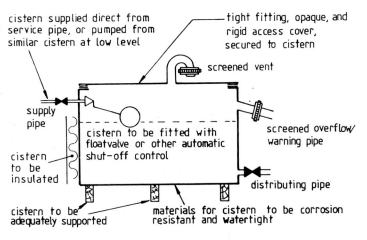

(a) General requirements

Figure 2.5   **Cistern requirements**

continued

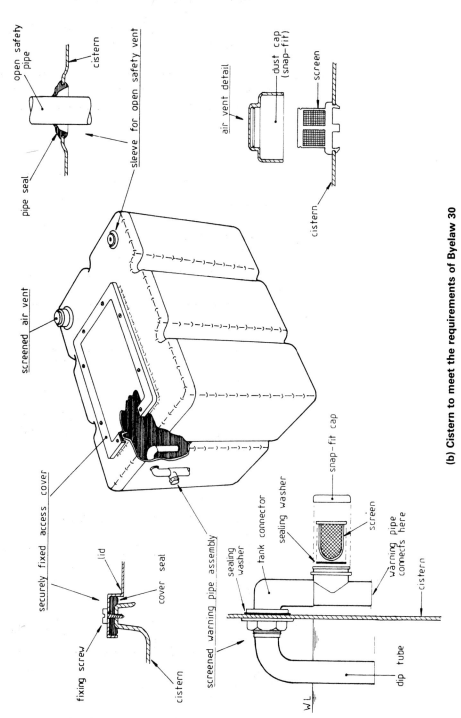

open safety pipe

cistern

pipe seal

sleeve for open safety vent

air vent detail

dust cap (snap-fit)

screen

cistern

screened air vent

securely fixed access cover

lid

cover seal

fixing screw

cistern

screened warning pipe assembly

sealing washer

tank connector

sealing washer

snap-fit cap

screen

warning pipe connects here

cistern

dip tube

WL

(b) Cistern to meet the requirements of Byelaw 30

Figure 2.5 **Cistern requirements**
continued

BS 6700 follows the Model Water Byelaws in specifying that water cisterns for domestic purposes must be of the 'protected' type, on the grounds that water is likely to be drunk from all taps in dwellings. This is a departure from past practice which should greatly improve cisterns hygienically.

All cisterns should be supported on a firm level base capable of withstanding the weight of the cistern when filled with water to the rim. Where cisterns are located in the roof space, the load should be spread over as many joists as possible (see figure 2.6).

(a) Flexible cisterns

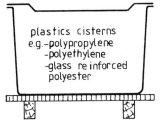

plastics cisterns
e.g.–polypropylene
–polyethylene
–glass reinforced
polyester

Continuous support needed over whole base area.

No connections to be made to base of plastics cisterns.

(b) Rigid cisterns

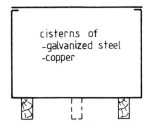

cisterns of
–galvanized steel
–copper

Two or more timber joists depending on size of cistern.

Continuous support not needed, and is undesirable for galvanized steel cisterns.

Figure 2.6   Support for cisterns

Occasionally large cisterns are buried or sunk in the ground. In these cases measures need to be taken to detect leakage and to protect the cistern from contamination (see figure 2.7).

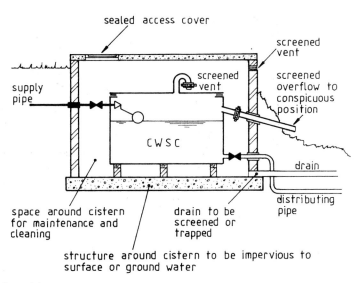

Figure 2.7   Sunken cistern

Cistern control valves

Every pipe supplying water to a cistern should be fitted with a float-operated valve or some other equally effective device to control the inflow of water and to maintain it at the required level.

Float valves should comply with BS 1212: Part 1, 2 or 3 (see figures 2.8 and 2.9) and be used with a float complying with BS 1968 or BS 2456 of the correct size corresponding to the length of the lever arm and the water supply pressure. Alternatively, float valves approved by the Water Research Centre may be used.

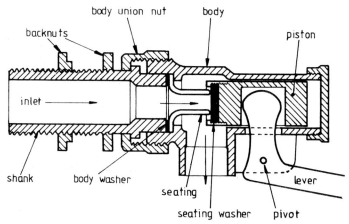

Portsmouth valves made to BS 1212: Part 1 are not suitable for WC cisterns, and if used in a storage cistern must have a double check valve arrangement fitted upstream.

Silencer tubes should *not* be fitted.

Figure 2.8   Portsmouth float valve to BS 1212: Part 1

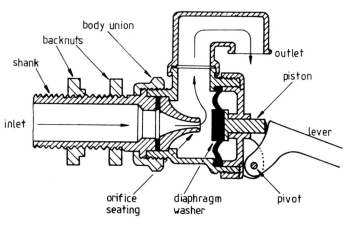

Figure 2.9   Diaphragm float valve to BS 1212: Part 2 (brass) and Part 3 (plastics)

Float valves should be clearly marked with the water pressure, temperature and other characteristics for which they are intended to be used.

Every float-operated valve should be securely fixed to the cistern and, where necessary, braced to prevent any movement of the float or cistern wall which might affect the inflow to the cistern, or cause noise.

Where a non-British Standard float-operated valve or other level control device is used, it must meet the requirements of the water byelaws and be approved by the Water Research Centre (see figure 2.10, 2.11 and 2.12).

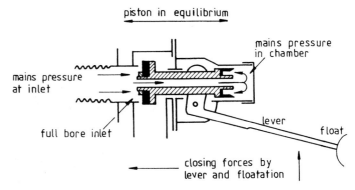

Figure 2.10   **Portsmouth equilibrium float valve**

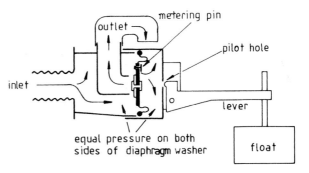

Figure 2.11   **Diaphragm equilibrium float valve**

Used to provide a full flow of water at all times, it completely eliminates 'dribble conditions' often associated with a conventional ball valve; this valve is particularly applicable to automatic pumping and booster systems and water treatment. 'Arclion' is a registered trade name of H. Warner & Son Ltd.

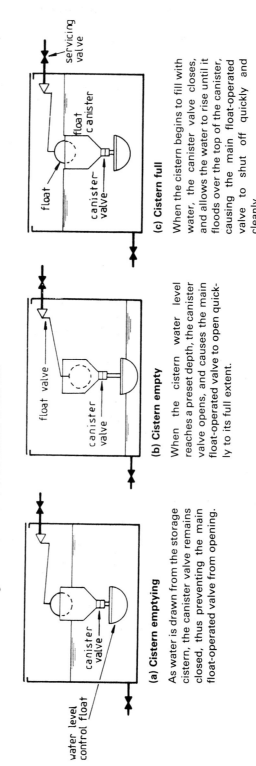

(a) Cistern emptying

As water is drawn from the storage cistern, the canister valve remains closed, thus preventing the main float-operated valve from opening.

(b) Cistern empty

When the cistern water level reaches a preset depth, the canister valve opens, and causes the main float-operated valve to open quickly to its full extent.

(c) Cistern full

When the cistern begins to fill with water, the canister valve closes, and allows the water to rise until it floods over the top of the canister, causing the main float-operated valve to shut off quickly and cleanly.

Figure 2.12   Arclion® Delayed action float valve

Connections to cisterns

These should be made as shown in figures 2.13 and 2.14.

Warning/overflow pipe must be large enough to carry away leakage under worst conditions.

Dimensions are in millimetres.

Cistern must comply with Byelaw 30 if it is used for domestic purposes.

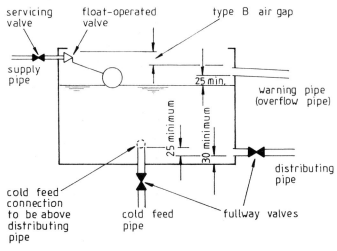

(a) Small cisterns

Washout pipe to be plugged when not in use and must discharge to open air at least 150 mm above any drain.

Overflow pipe must be large enough to carry away leakage under worst conditions.

Cold feed and cold distributing pipes to be fitted with corrosion resistant strainers.

Cold feed pipe to be above distributing pipe.

Dimensions are in millimetres.

Cistern must comply with Byelaw 30 if it is used for domestic purposes.

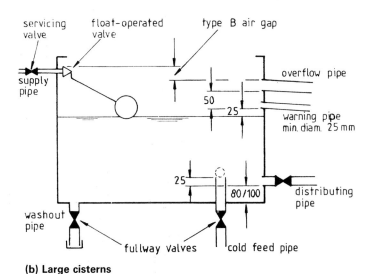

(b) Large cisterns

Figure 2.13   Connections to cisterns

Linked cisterns

On occasion, to provide large quantities of water storage, or because of space restrictions, two or more cold water storage cisterns may need to be linked. Figure 2.14 shows two methods which permit cisterns to be cleaned, repaired or replaced without interruption of supplies to the building.

To avoid *Legionella* cisterns should:
- be small enough to ensure a rapid turnover and thus prevent stagnation.
- have float valves arranged to open and close together.
- have inlet and outlet connections at opposite ends.
- be regularly inspected and maintained in clean condition.
- conform to the requirements of Byelaw 30.

The diagram shows separate overflow pipes. A common overflow pipe or warning pipe may be fitted provided cisterns are linked to form one storage unit

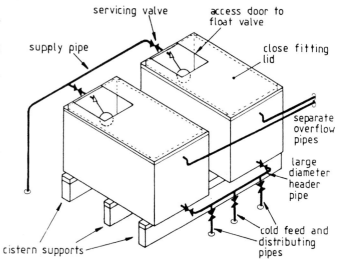

(a) Cisterns connected in parallel

This method is preferred for prevention of *Legionella*.

To take cistern 1 out of commission for cleansing:
- fit temporary connection between link (b) to distributing pipe;
- remove link (a) and cap off close to cisterns.

To take cistern 2 out of commission for cleansing:
- fit temporary connection to float valve in cistern 1 and disconnect branch pipe to float valve in cistern 2;
- remove links (a) and (b) and cap off near cisterns.

Sterilise any pipes and fittings used before they are fitted.

Sterilise cistern and pipes before putting back into service.

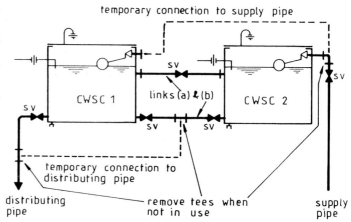

(b) Cisterns connected in series

Figure 2.14   Linked cisterns

Cisterns mounted outside buildings

Whether fixed to the building itself or supported on an independent structure, cisterns outside buildings, see figure 2.15, should be enclosed in a well-ventilated, yet draught-proof housing. This should be constructed to prevent the entry of birds, animals and insects, but provide access to the interior of the cistern for maintenance. Ventilation openings should be screened by a corrosion-resistant mesh with a maximum aperture size of 0.65 mm.

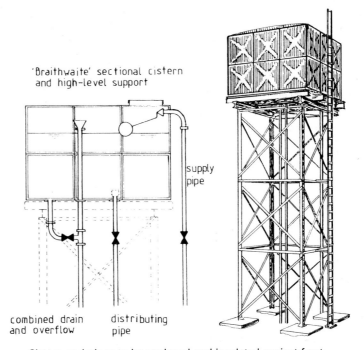

'Braithwaite' sectional cistern and high-level support

supply pipe

combined drain and overflow   distributing pipe

Cistern and pipes to be enclosed and insulated against frost.

Enclosure to be ventilated, but draughtproof, and arranged to prevent entry by birds, animals and insects.

Overflow/drain pipe must terminate in a conspicuous position above ground level.

Figure 2.15   Typical exterior storage cistern

Large cisterns

Generally these provide over 5000 l storage and are often made up of preformed panels or are constructed of concrete. They should preferably be divided into two or more compartments to avoid interruption of the water supply when carrying out repairs or maintenance to the cistern.

Cistern capacities

Clause 5.3.3 of BS 6700 makes the following recommendations for houses:

smaller houses	cold water only	– 100 l to 150 l
	hot and cold outlets	– 200 l to 300 l
larger houses	per bedroom	– 100 l

See also section 5.3.

Pipework to and from cisterns

Supply and distributing pipes to and from cisterns should comply with the following recommendations.

(1) All pipework should be insulated to reduce heat losses and gains, to minimise frost damage, and to prevent condensation.
(2) Pipes should preferably be laid to a fall to reduce the risk of air locks and to facilitate filling and draining.
(3) Pipes should be closely grouped for neatness, but not so close as to gain heat from one another.
(4) Pipes should be securely fixed and adequately supported.
(5) Only flush pipes fed from a flushing cistern or trough shall deliver water to WCs and urinals.
(6) A cold feed pipe should not be used other than to supply the hot water apparatus for which it is intended.

Warning and overflow pipes

Warning and overflow pipes (see figure 2.16) serve two purposes:

(1) to give warning that inlet valve to cistern has failed to close,
(2) to remove safely from the buildings any water which does leak from the inflow pipe.

On cisterns of less than 1000 l one pipe only will serve both as an overflow pipe and a warning pipe, but on larger cisterns it may be necessary to have a separate pipe for each function (see figures 2.16–2.18).

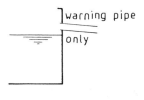

(a) Small cisterns

Small cisterns of up to 1000 l nominal capacity must be fitted with a warning pipe and no other overflow pipe.

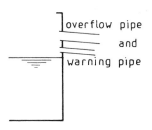

(b) Medium cisterns

Cisterns of between 1000 l and 5000 l nominal capacity must be fitted with an overflow pipe and a warning pipe.

Cisterns of more than 5000 l nominal capacity may have other warning devices fitted in place of the warning pipe (see figures 2.17 and 2.18).

Figure 2.16   **Warning and overflow pipes for cisterns**

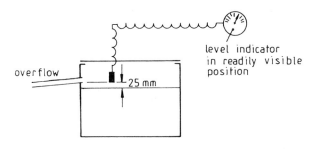

(a) Electrically operated warning device

Cisterns of between 5000 l and 10 000 l nominal capacity must be fitted with an overflow pipe and a warning pipe. Alternatively the warning pipe may be replaced by a level indicator that will clearly show when the water level is 25 mm below the overflowing level.

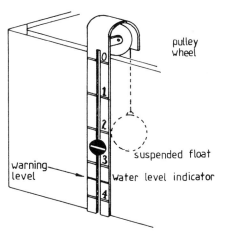

(b) Float-operated warning device

For use with cisterns of between 5000 l and 10 000 l nominal capacity

Figure 2.17   **Overflow and warning arrangements for large cisterns**

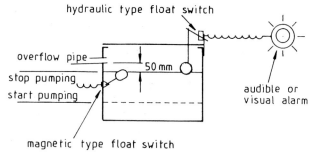

(a) Electrically operated alarm

Cisterns of more than 10 000 l nominal capacity must be fitted with an overflow pipe and a warning pipe. Alternatively the warning pipe may be replaced by an audible or visual alarm that will clearly show when the water level is 50 mm below the overflowing level.

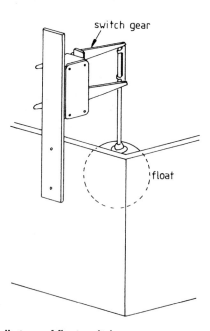

(b) Hydraulic type of float switch

Float switch can be used to operate an audible or visual alarm. Also used for water level control.

Figure 2.18   Overflow and warning arrangements for very large cisterns

The following recommendations for warning pipes and overflow pipes are noted (see figures 2.19–2.21).

(1) Overflow and warning pipes should be made of rigid, corrosion-resistant material.

(2) Overflow and warning pipes fitted to feed and expansion cisterns must be of metal, cross-linked polyethylene, polybutylene or chlorinated PVC, and be able to resist heat.

(3) The overflow pipe or pipes must be able to carry away all water which is discharged into the cistern in the event of the inlet control becoming defective, without the water level reaching the spill-over level of the cistern, or submerging the discharge orifice of the inlet pipe or valve.

(4) Warning and overflow pipes must fall continuously to points of discharge.

(5) Where overflow and warning pipes discharge through the external wall of a building they should be arranged to prevent cold draughts by turning down the warning pipe into the cistern and below the water line (see figure 2.19).

(6) Every warning pipe shall discharge in a conspicuous position, preferably outside the building where this is appropriate. In some circumstances warning pipes may be permitted to terminate into a special bath overflow fitting.

(7) No warning pipe must discharge into a WC pan or flush pipe.

(8) Warning pipes from feed and expansion cisterns must be separate from those from storage cisterns.

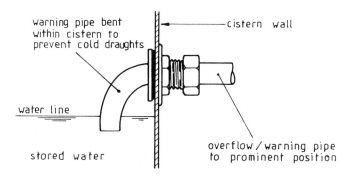

Figure 2.19   **Overflow/warning pipe**

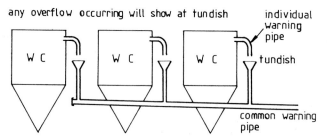

any overflow occurring will show at tundish

Cisterns must all be at the same level.

Individual warning pipes must terminate over tundish before connection to common pipe.

Common warning pipe must terminate in a conspicuous position.

Figure 2.20   **Common warning pipes to WCs**

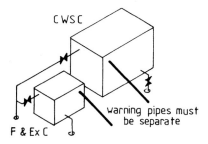

Figure 2.21   **Warning pipes from feed and expansion cisterns**

Warning pipes can be combined into one outlet provided that:

- the outlet is labelled to indicate the cisterns served, for quick location of the fault;
- individual warning pipes from WC cisterns empty into a tundish so that any discharge is readily visible (see figure 2.20).

2.4   Valves and controls

Supply stopvalves

On service pipes these must comply with the following British Standards (see figure 2.22):

> Up to 50 mm diameter
>   above ground to BS 1010: Part 2
>   below ground to BS 5433 or BS 2580
> 50 mm diameter or more
>   flanged gate valve to BS 5163

Stopvalves to BS 1010 are for above ground use only.

Stopvalves to BS 5433 for underground use are of heavier quality and made of corrosion resistant material such as gun metal or DZR brass.

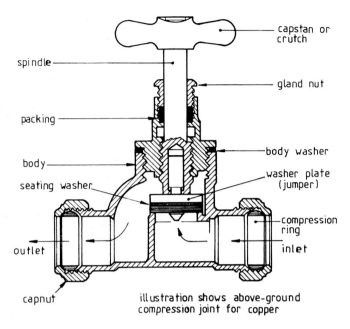

illustration shows above-ground compression joint for copper

(a) Screwdown stopvalve to BS 1010 and BS 5433

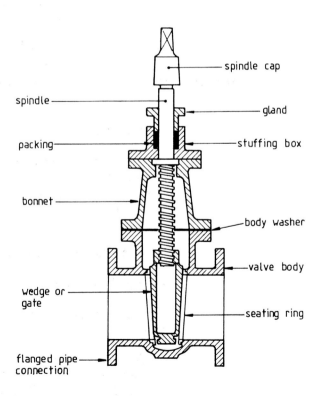

(b) Flanged gate valve to BS 5163

Figure 2.22   **Stopvalves**

A stopvalve must be provided on the service pipe, inside the building, to control the whole of the water supply to the building. It should be in an accessible position above floor level, and near the point of entry of the pipe.

When a stopvalve is installed on an underground pipe it should be enclosed in a pipe guard under a surface box.

Stopvalve positions outside buildings

Service pipes to buildings can be arranged in numerous different ways as shown in figures 2.23–2.25. In each case separate dwellings supplied must have a controlling stopvalve in a position that will allow the supply to be turned off in an emergency without affecting any other property. Additionally, water undertakers require that each dwelling has a stopvalve which is readily accessible to them in an emergency.

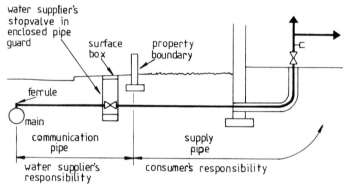

(a) Individual dwelling

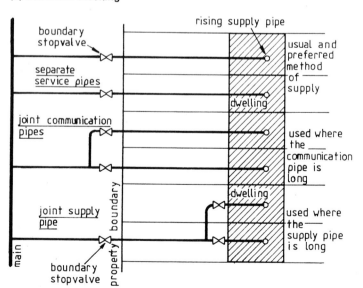

Note Joint supply pipes often present problems when repairs have to be carried out, especially where a number of dwellings are supplied. For this reason, joint supply pipes are not favoured by water authorities, who will usually insist on separate service pipes.

(b) Methods of supplying more than one dwelling

Figure 2.23   Stopvalves to dwellings

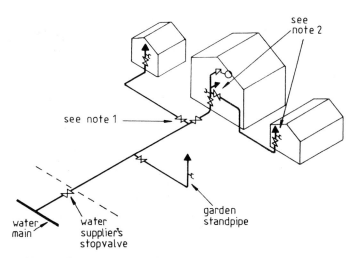

Note 1 Separate buildings supplied require separate controlling stopvalves outside the buildings to control whole of branch supply.

Note 2 Where one building is supplied from another, a stopvalve should be fitted to control the whole of the branch pipe in addition to one at the point of entry to the building.

Figure 2.24   Stopvalves in various positions within a single property

(a) Internally separated system

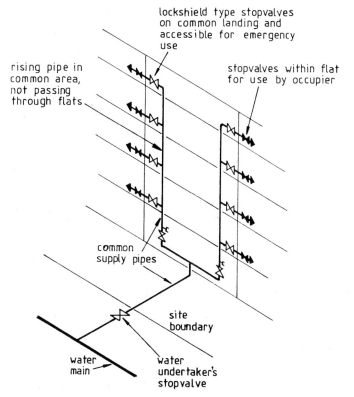

Figure 2.25   Common supply pipes to flats

continued

(b) Externally separated system

The externally separated system is the method preferred by water authorities who may also insist on a separate mains connection for each flat.

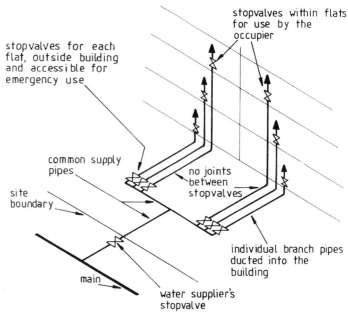

stopvalves for each flat, outside building and accessible for emergency use

stopvalves within flats for use by the occupier

common supply pipes

no joints between stopvalves

site boundary

individual branch pipes ducted into the building

main

water supplier's stopvalve

Figure 2.25 continued   **Common supply pipes to flats**

Servicing valves

Servicing valves must be provided in accessible positions to enable the flow of water to individual or groups of appliances to be controlled and to limit inconvenience caused during maintenance and repair (see figures 2.26 and 2.27).

(a) Supply stopvalves

Supply stopvalves that meet the requirements of Byelaw 64.

Up to 50 mm diameter:
  ○ screwdown stopvalves to BS 5433
  ○ plugcock to BS 2580
  ○ screwdown stopvalves to BS 1010 (above ground use only)

Above 50 mm diameter:
  ○ flanged gatevalve to BS 5163

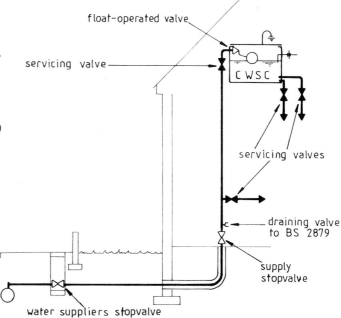

(b) Servicing valves

Note WC cistern not to be fitted with a lever-operated valve.

Key to use of servicing valves:

① Screwdriver-operated, slot type spherical plug valve to BS 6675
② Lever-operated spherical plug valve to BS 6675
③ Wheel-operated valve (gatevalve) to BS 5154
④ Any supply stopvalve that meets the requirements of byelaw 64

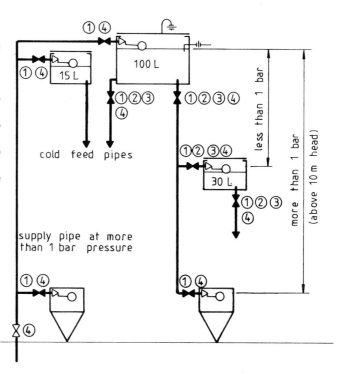

Figure 2.26   Control valves within dwellings

(a) Gatevalve to BS 5154

Not for use above 1 bar pressure

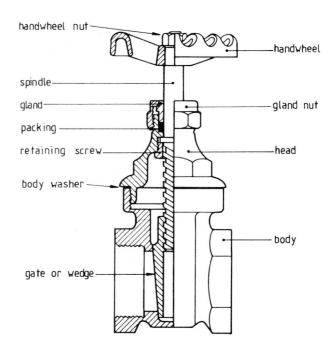

(b) Spherical plug valves

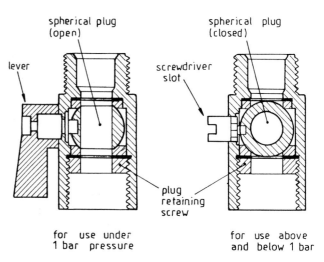

Figure 2.27 **Types of servicing valve**

Servicing valves must be fitted:

- immediately before every float valve connected to a service pipe,
- on every cold feed and distributing pipe from any feed cistern or storage cistern of more than 18 l capacity,
- to every pipe carrying water from a hot water storage cylinder, tank or cistern.

Servicing valves should preferably not be fitted on the cold feed pipe from a feed and expansion cistern to a primary hot water circuit (see section 4.2).

2.5 Water revenue meters

This section should be regarded as for information only. It is probable that regulations will be made as to water metering matters in due course.

Meters may be fitted by the consumer's contractor or by the water undertaker. Where fitting is done by a private contractor the water supplier will need to be consulted and agreement reached on siting and installation details.

The revenue meter will be supplied by and will remain the property of the water supplier.

The preferred position for the meter is at the boundary of the premises at the end of the communication pipe so that it will register the whole supply (see figure 2.28).

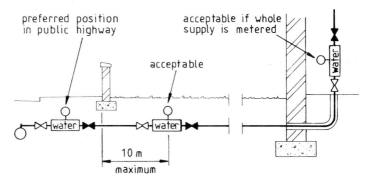

Figure 2.28   Meter positions

In the case of premises with multiple occupations, e.g. flats, and where underground installations are not practicable, an internal installation may be acceptable provided the whole supply is registered.

Notes on fitting

The following points on fitting should be noted:

- The meter installation should comply with BS 5728.
- The meter must be protected from risk of damage by shock or vibration.
- The method of connection should permit meter changes without the use of heat or major disturbance of pipework.
- Connector unions should have flat seats and be complete with 1.5 mm thick washers to allow for tolerance on meter lengths.

- Meter lengths – G3/4B meters, 134 mm;
  – G1B meters, see BS 5728.

Flow rates

Nominal flow rate for the G3/4B meter is 1.5 m³/h, and for the G1B meter is 2.5 m³/h.

Continuity bonding

The Institution of Electrical Engineers' Regulations for Electrical Installations do not allow the use of water pipes as an electrode for earthing purposes. However, any metal water supply pipe must be bonded to the electrical installation main earth terminal as near as possible to its point of entry into the building.

A suitable conductor should be installed between pipes on either side of the meter and stopvalves, to protect the installer against electrical fault, and for the maintenance of the earth connection during use, and particularly when the meter is being replaced.

External meter installations

- The chamber should be well constructed of brick, concrete, GRP or PVC, with a cover marked 'Water meter' and fitted with slots or lifting eyes. Covers should *not* be of concrete.
- Pipes, cables and drains other than meter pipework must not pass through the meter chamber.
- A meter below ground should be installed in the horizontal position to facilitate meter reading.

See table 2.1 and figure 2.29.

Table 2.1   **Recommended internal dimensions for meter chambers** (dimensions in mm)

Size of meter	Size of chamber			Remarks
	Length	Breadth	Depth	
15 to 20	430	280	To suit	Or 380 circular
25	600	600	pattern	
40 to 50	900	600	of meter	
80*	1900	750	but not	
100*	2000	750	less than	
150*	2150	750	750	

* Dimensions for these sizes take account of compound assemblies but do not provide for isolating valves.

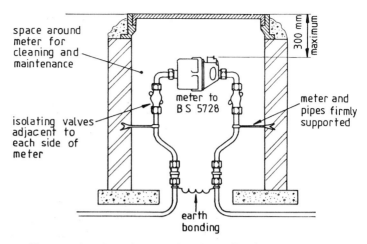

Pipes passing through meter chamber wall or floor should have clearance space to permit changing of meter and to prevent damage through ground movement.

Stopvalves shown are not in the most accessible positions.

Figure 2.29   **Below ground meter installation**

Note   Whilst BS 6700 and many Water Authorities favour raising the meter as shown in figure 2.29 to make meter reading and changing easier, I would prefer the meter to remain at the same level as the service pipe, as in figure 2.30, in order to comply with the water byelaw requirement of 750 mm depth to avoid frost damage.

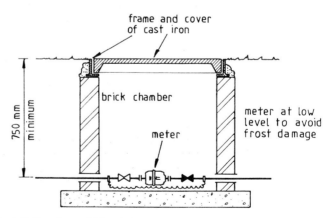

Figure 2.30   **Meter installation to avoid frost damage**

Internal meters

Internal meters may be fixed horizontally or vertically provided the dial is not more than 1.5 m above floor level and readily visible for reading (see figure 2.31).

(a) General requirements

Pipework to be adequately supported.

Meter to be protected from frost, especially in exposed positions, e.g. garage.

Meter in cupboard to be brought forward to within 300 mm of the front of cupboard.

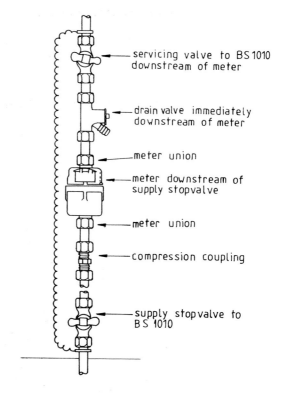

servicing valve to BS 1010 downstream of meter

drain valve immediately downstream of meter

meter union

meter downstream of supply stopvalve

meter union

compression coupling

supply stopvalve to BS 1010

(b) Positioning

Meter in a cupboard may be brought forward for ease of reading provided it is properly supported.

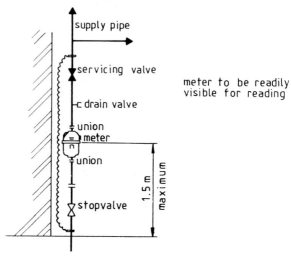

supply pipe

servicing valve

drain valve

union

meter

union

stopvalve

1.5 m maximum

meter to be readily visible for reading

Figure 2.31   Above ground meter installation

Non-revenue meters

Installation should be as described for revenue meters but the water undertaker will not need to be consulted about the meter's position.

Patent meter connections

These save space and installation costs (see figure 2.32).

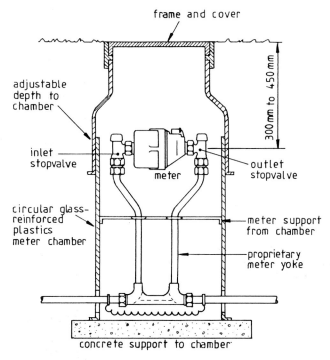

(a) Below ground, using meter yoke

Figure 2.32   **Patent meter connections**

continued

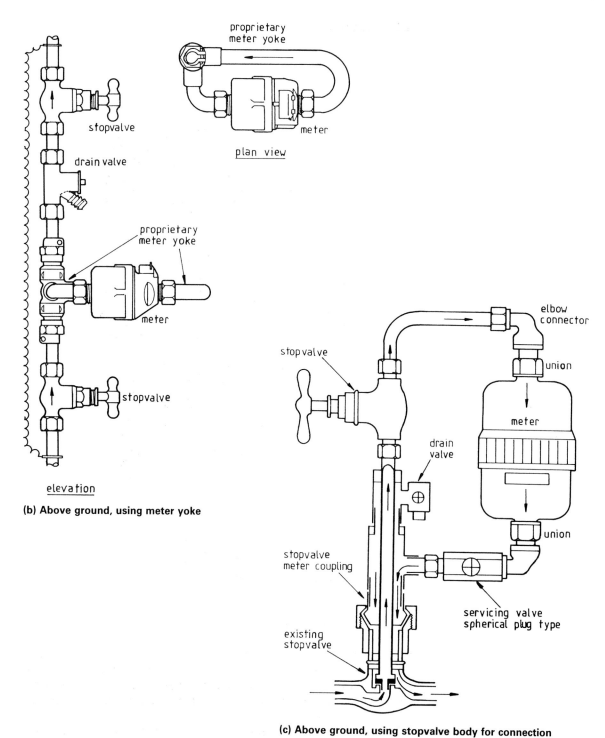

(b) **Above ground, using meter yoke**

(c) **Above ground, using stopvalve body for connection**

Figure 2.32 **Patent meter connections**
continued

Meter readings

A typical domestic meter reading is shown in figure 2.33.

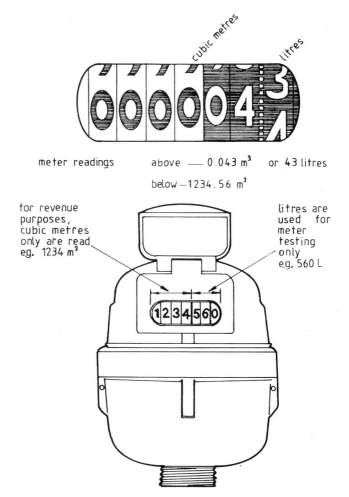

Figure 2.33   Meter reading – typical domestic meter

2.6 Boosted systems

Water supplies to buildings vary greatly in pressure and quantity available. In some cases this may give rise to intermittent supplies and in others, especially high-rise developments, parts of the building may be above the pressure limit of the mains supply. In these situations there is a case for the use of pumps.

There are two ways that pumps can be used to deal with the above problems. These are by direct boosting and indirect boosting. Indirect systems are more common than direct systems; the latter are rarely permitted by water suppliers because they reduce the mains pressure available to other consumers and can increase the risks of backsiphonage. Under no circumstances should any pump be connected directly to a pipe without first obtaining the written consent of the water undertaker.

Booster pumps can cause excessive aeration. Although this does not cause deterioration of water quality, the 'milky' appearance can cause concern amongst consumers.

Basic systems

The basic systems are as follows:

- simple direct boosting, see figure 2.34,
- direct boosting to header and duplicate storage cisterns, see figure 2.35,
- indirect boosting to storage cistern, see figure 2.36,
- indirect boosting with pressure vessel, see figure 2.37.

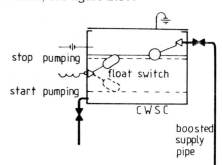

When used for drinking water, the storage vessel should be of the 'protected' type.

Pump control provided by level switch or similar device in the high-level storage cistern.

Pumps switch on when the level of water drops to a predetermined depth (normally about half the depth of the cistern) and they should switch off when the water level rises to about 50 mm below the shut-off level of the float-operated valve.

The frequency with which the pumps switch on and off should be limited to reduce wear on them, but the frequency of operation depends on the quantity of water used and stored, and on the pump rating.

Where the water supplier permits, pumps are connected to the incoming supply pipe to enable the pressure head to be increased.

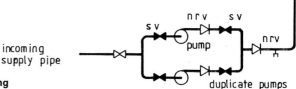

Figure 2.34   **Simple direct boosting**

System used for large and high-rise buildings.

Cisterns at high-level supply non-drinking water.

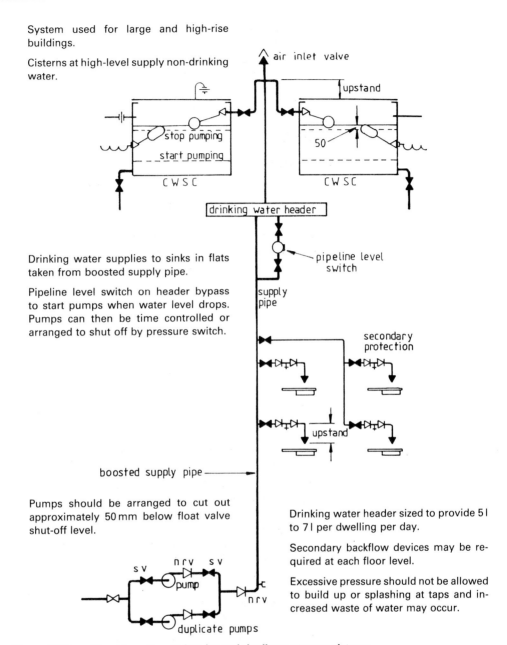

Drinking water supplies to sinks in flats taken from boosted supply pipe.

Pipeline level switch on header bypass to start pumps when water level drops. Pumps can then be time controlled or arranged to shut off by pressure switch.

Pumps should be arranged to cut out approximately 50 mm below float valve shut-off level.

Drinking water header sized to provide 5 l to 7 l per dwelling per day.

Secondary backflow devices may be required at each floor level.

Excessive pressure should not be allowed to build up or splashing at taps and increased waste of water may occur.

Figure 2.35   Direct boosting to header and duplicate storage cisterns

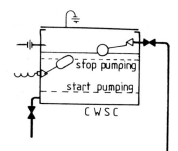

Float switch in storage cistern to operate pumps as water level in cistern rises and falls.

Pumps should cut out as water level rises to approximately 50 mm below float valve shut-off level.

For drinking water, protected cisterns should be used.

Break cistern should have effective capacity equivalent to at least 15 minutes pump output.

Float switch in break cistern to shut off pumps when water level drops to approximately 225 mm above pump suction connection. This will ensure that pump does not run dry.

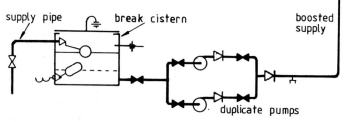

Figure 2.36   **Indirect boosting to storage cistern**

For use in buildings where a number of storage cisterns are supplied at various levels and where it is not practicable to control pumps by level switches.

Normally pressure vessel, pumps, air compressor and control equipment are purchased as a packaged pressure set.

Floors above limit of mains pressure supplied via break cistern and pumps.

Unboosted supply to floors within limit of mains pressure.

Drinking water cisterns in dwellings must be protected to Byelaw 30 standards.

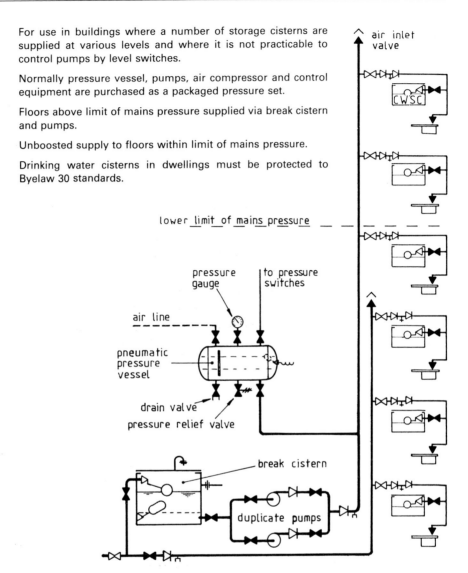

Secondary backflow protection is required at each floor level where separately chargeable premises are supplied.

Figure 2.37   Indirect boosting with pressure vessel

The *pneumatic pressure vessel*, as seen in figure 2.38, contains both compressed air and water. As the water is drawn off, the water level drops, the air expands with a resulting loss of water, and a float switch starts the pumps. The water level rises to a predetermined level at which the float switch stops the pumps. The sequence of events will continue to supply the building until the water level falls and the cycle begins again. As and when air is lost from the pressure vessel by absorption into the water, pressure will fall, and the air is replaced via the air compressor.

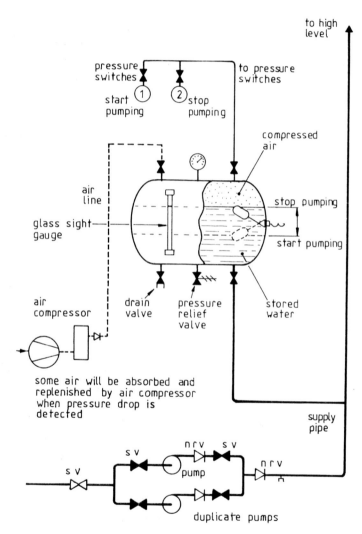

Figure 2.38   Pneumatic pressure vessel

Some systems incorporate a sealed pressure vessel containing a flexible membrane to separate the air from the water. This will reduce air loss and eliminate the need for a permanent compressor. However, air pressures should be checked at intervals and replenished if necessary. Some units contain nitrogen instead of air, and it is important that these are topped up with nitrogen, not air or some other gas (see figure 2.39).

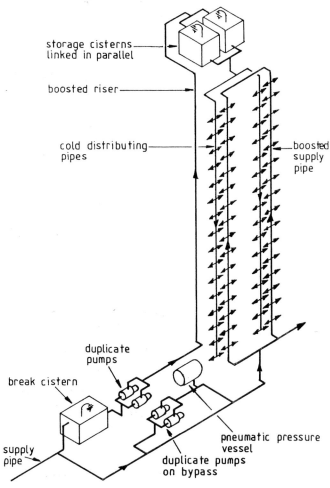

storage cisterns linked in parallel

boosted riser

cold distributing pipes

boosted supply pipe

duplicate pumps

break cistern

pneumatic pressure vessel

supply pipe

duplicate pumps on bypass

Drinking water supply direct from mains using pneumatic pressure vessel.

Indirect boosting for storage cisterns.

Control valves and backflow prevention devices omitted for clarity.

Figure 2.39   **Typical installation to multi-storey building**

Pumps and equipment

Pumps and other associated equipment are usually located within the building being served, preferably as near as possible to the point of entry of the incoming pipe.

Electrically driven centrifugal pumping plant is normally used; pumps and other equipment should be duplicated.

Pumps may be either horizontal or vertical types, directly coupled to their electric motors. A solid foundation is essential for all motors and pumps and anti-vibration mountings should normally be specified.

Automatic control of pumping plant is essential and pressure switches, level switches, or high-level and low-level electrodes should give reliable control. Other methods of control, both mechanical and electrical, could be considered. Pumping should be controlled using a pump selector switch and an ON/OFF/AUTO control. Motor starters should incorporate overload protection.

Pumps should be installed in duplicate and sized so that each pump is capable of overcoming the static lift plus the friction losses in both pipework and valves. Where pumps are connected directly to the service pipe, allowance should be made for the minimum operating pressure in the service pipe, since the pump head is added to this and does not cancel out any existing pressure.

Care should be taken in pump and pipe sizing to minimise the risk of water hammer due to surge when pumps are started and stopped.

Transmission of pump and motor noise via pipework can be reduced by the use of flexible connections. Small-power motors of the squirrel cage induction type are suitable for most installations. Low-speed pumps are preferable to promote a long efficient life and reasonably quiet operation. The fitting of motors with sleeve type super-silent bearings should be considered for quiet running.

All pipework connections to and from pumps should be adequately supported and anchored against thrust to avoid stress on pump casings and to ensure proper alignment.

Most small air compressors used for charging pneumatic pressure vessels are of the reciprocating type, either air-cooled or water-cooled. A water-cooled after-cooler for the condensation and extraction of oil and moisture from the compressed air should be installed. The air to be compressed should be drawn from a clean cool source and should be protected from contamination. Check or non-return valves should offer a minimum of frictional loss and should be non-concussive.

The pump room (see figure 2.40) should be of adequate size to accommodate all the plant and also provide adequate space for maintenance and replacement of parts; it should be dry, ventilated and protected from frost and flooding. Entry of birds and small animals must be prevented. Access should be restricted to authorised persons.

Provision should be made for the pumps to be supplied by an alternative electricity supply in the event of mains failure.

Maintenance and inspection

A responsible person should be appointed to oversee the proper installation of any scheme, and the user should arrange for regular maintenance and inspection of the pumps and plant.

All work carried out and inspections made should be recorded in a suitable log book which should be kept in the plant room.

2.7 Water softeners

The purpose of water softeners is to reduce scale formation in hot water systems and components (see figure 2.41).

Advantages

Some of the advantages of water softeners are as follows:

(1) Savings in soap and reduction of scum, resulting in:
   - reduced expenditure on soap purchase (small savings);
   - easier cleaning of appliances;
   - cleaner crockery from dishwashers.
(2) Smooth,gentle feel of bath and shower water.
(3) Scale reduction in appliances and components, resulting in:
   - longer life for cylinders, immersion heaters and other components;
   - less maintenance, e.g. shower outlets less likely to become clogged with fur.

Disadvantages

(1) Additional installation costs (manufacturers claim a six-year pay back).
(2) Cost of running the unit, e.g. electricity supply and salt.
(3) Drain needed for brine rinse.
(4) User must add salt periodically.

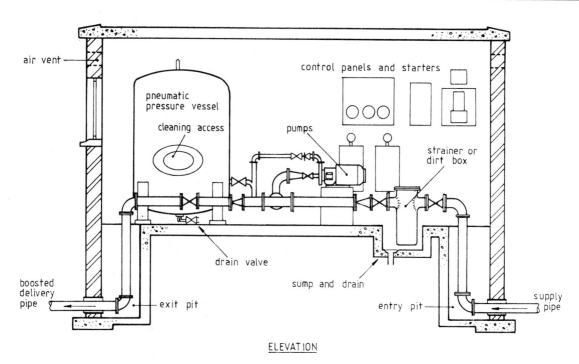

ELEVATION

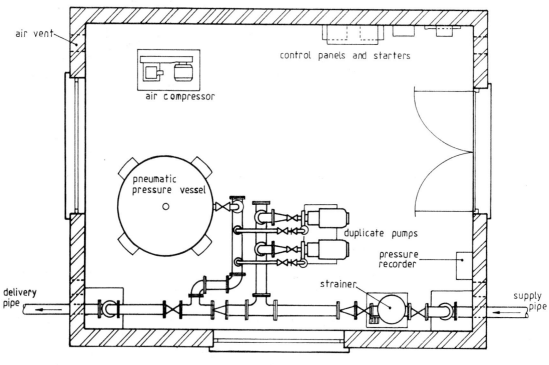

PLAN

Figure 2.40   Typical pump room

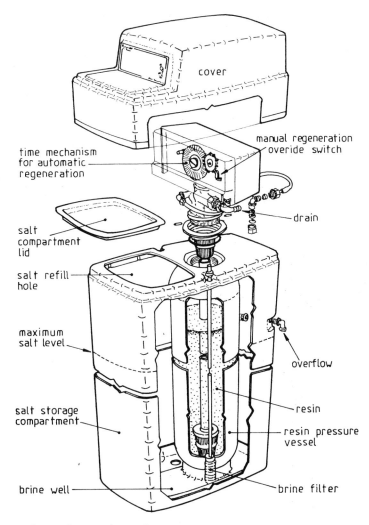

labels:
cover

manual regeneration
overide switch

time mechanism
for automatic
regeneration

drain

salt
compartment
lid

salt refill
hole

maximum
salt level

overflow

salt storage
compartment

resin

resin pressure
vessel

brine well

brine filter

Figure 2.41   **Water softener, base exchange type**

Installation

Water softeners should be sited near the incoming supply pipe and where drain access is available (see figure 2.42).

Electricity supply is required for automatic control and operation.

Water softeners of the salt regenerated type installed in dwellings are considered to be a minor backflow risk and as such a check valve should be installed on the supply pipe before the softener connection.

Drinking water supply should be taken off before the softener, and upstream of check valves.

A softener located other than in a dwelling requires the use of a double check valve arrangement or a combined check and anti-vacuum valve.

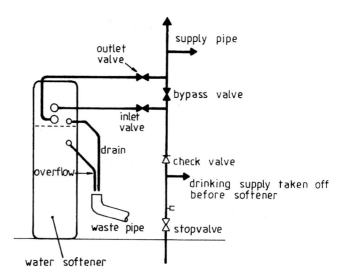

Figure 2.42   **Installation of base exchange water softener**

Other forms of treatment

BS 6700 deals briefly with water softeners, but does not refer to other forms of treatment, which are shown in figures 2.43 and 2.44. It is not the intention here to recommend their use, rather to point out that there are such treatments readily available, and that they can in certain circumstances have a limited application.

Before any of the chemical devices are used, advice should be sought from local water undertakers, particularly regarding toxicity.

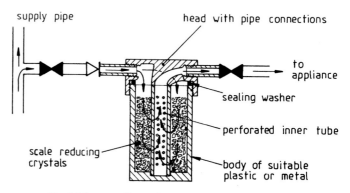

Useful for supplies to instantaneous water heaters.

Figure 2.43   **Pipeline dispenser**

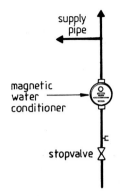

Precipitates the hardness salts into microscopic crystals by passing the water through a magnetic field.

Prevents scale forming as long as water is moving, but effects are said to lessen in stored water.

Figure 2.44   Magnetic water conditioner

Chapter 3
Hot water supply

The following factors should be considered in the selection and the design of hot water supply systems:

(1) quantity of hot water required;
(2) temperature in storage and at outlets;
(3) cost of installation and maintenance;
(4) fuel energy requirements and running costs;
(5) waste of water and energy.

Hot water supply cannot be considered in isolation from central heating because systems commonly combine both functions. This book, like BS 6700, is primarily concerned with hot water supply and refers only to central heating in combined hot water and heating systems up to 44 kW output.

3.1   System choice

There are many methods by which hot water can be supplied, ranging from a simple gas or electric single point arrangement for one outlet, to the more complex centralised boiler systems supplying hot water to a number of outlets.

Figure 3.1, adapted from BS 6700, sets out a number of ways of supplying hot water, many of which are described in the following pages.

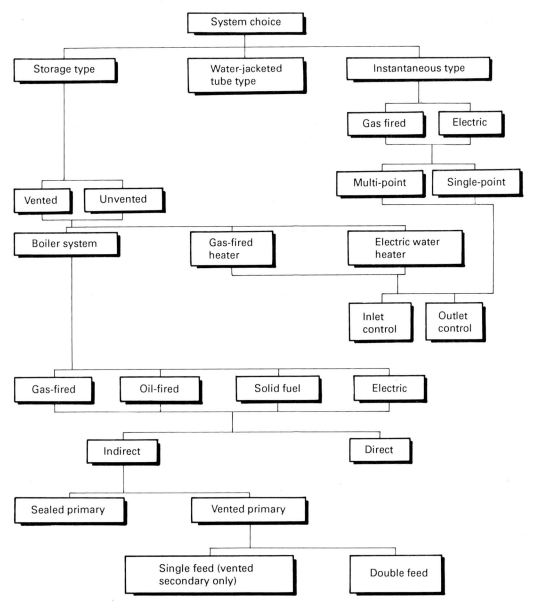

The system chosen depends on circumstances and the needs of the user, and may require the use of one method or a combination of two or more.

Figure 3.1   **Alternatives for hot water systems**

3.2   Instantaneous water heaters

Instantaneous water heaters (see figures 3.2 to 3.6) should be chosen with the following considerations in mind.

(1) Some of these heaters have relatively high power ratings (up to 28 kW if gas fired) so it is important that adequate gas or electricity supplies are available.

(2) The water in instantaneous water heaters is usually heated by about 55°C at its lowest flow rate, and its temperature will rise and fall inversely to its flow rate.

(3) Where constant flow temperature is important, the heater should be fitted with a water governor at its inflow. Close control of temperature is of particular importance for showers.

(4) To attain constant temperatures on delivery, water flow and pressure must also be constant. Variations in pressure can cause flow and temperature problems when the heater is in use, and when setting up or adjusting flow controls.

(5) The use of multi-point heaters for showers should be avoided, except where the heater only feeds a bath with a shower over it.

(6) Gas-fired heaters fitted in bathrooms must be of the room-sealed type. Room-sealed types are preferred in other locations.

(7) Electrically powered heaters in bathrooms must be protected against the effects of steam and comply with the Regulations for Electrical Installations of the Institution of Electrical Engineers.

Position heater near most-used appliance, usually the kitchen sink.

Where pipe runs from a multi-point heater are likely to be long consider using a number of single point heaters.

Multi-point heaters operate most satisfactorily when only one outlet is used at any one time.

Flow rate is variable and heater is generally not suited for use with a shower.

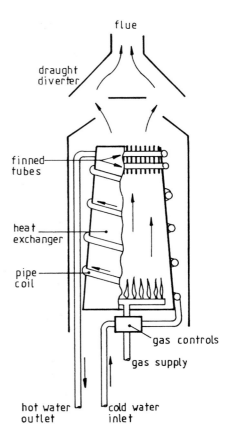

Figure 3.2   Typical instantaneous water heater, gas-fired with conventional flue

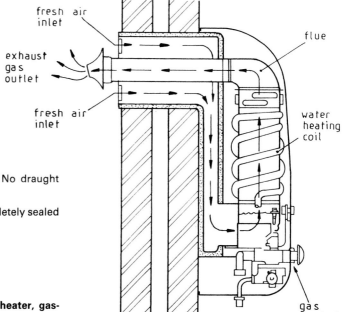

Flue inlet and outlet at equal air pressure. No draught problems to affect flame.

Combustion chamber, air inlet and flue completely sealed from room.

Figure 3.3   Typical instantaneous water heater, gas-fired with balanced flue

(a) Directly supplied heater

Constant flow rate needed to maintain 55°C temperature difference between feed water and heated water.

Pressure and flow variations will affect temperatures at outlets.

Showers are not recommended because of possible loss of constant temperature control and pressure.

Use only thermostatically controlled shower mixer.

The usual arrangement is direct from the supply pipe as shown here because installation cost is lower. However, supply from storage will give constant flow.

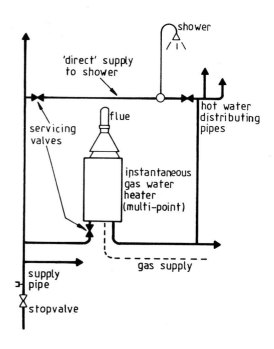

(b) Indirectly supplied heater

High installation cost compared with mains-fed system

Constant pressure from storage for shower and other fittings gives more stable temperature control.

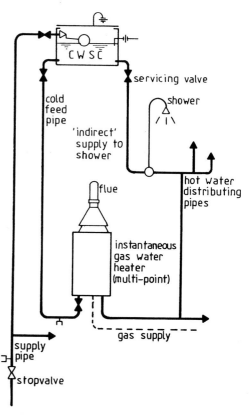

Figure 3.4   Centralised gas-fired instantaneous water heater installations

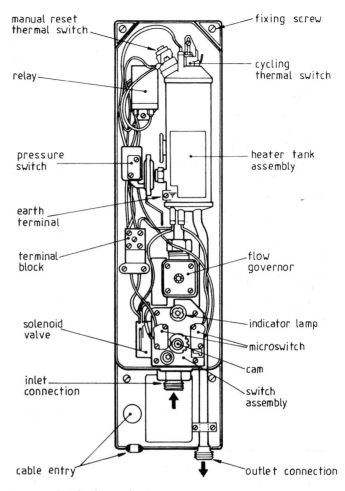

Figure 3.5 **Instantaneous electric shower heater**

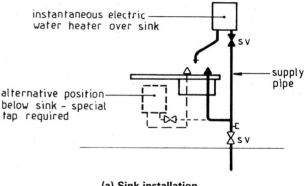

instantaneous electric
water heater over sink

s v

supply
pipe

alternative position
below sink – special
tap required

s v

(a) Sink installation

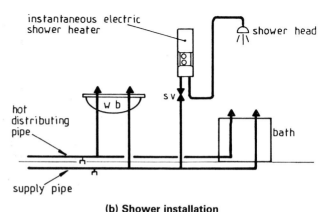

instantaneous electric
shower heater

shower head

hot
distributing
pipe

w b

s v

bath

supply pipe

(b) Shower installation

Where shower unit is fitted it will need a minimum head of
10.5 m.

Flexible shower outlet may be a contamination risk if nozzle can
become submerged (see figure 3.49 (b)).

Figure 3.6   Typical uses for instantaneous electric water heaters

3.3 Water-jacketed tube heaters

The water drawn off for use passes through a heat exchanger in a
reservoir of primary hot water (see figure 3.7). The size of this reservoir
and its heat input determines the volume and rate of flow of hot water that
can be provided without an unacceptable temperature drop. The cold
water feed to the heater may be from the mains or from a storage cistern.

Water-jacketed tube heater is a form of instantaneous heater.

Primary circuit may be vented system or sealed system.

Heat exchanger warms secondary supply water as it passes through.

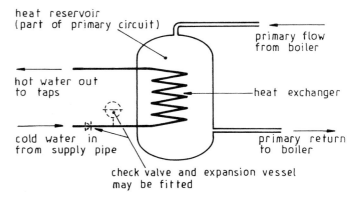

(a) Basic principles

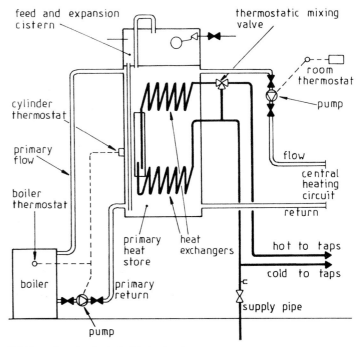

(b) Use of water-jacketed tube heater

Note 1 Primary water from the boiler flows to the heat store as programmed by the cylinder thermostat. Hot water is pumped to the radiator heating circuits and returns to the heat store. Cooler water from the heat store then returns to be reheated in the boiler.

Note 2 Cold water under mains pressure from the supply pipe enters the lower heat exchanger to be partially heated. It then passes through the upper heat exchanger where it is fully heated before being distributed to the taps.

Drawing shows the 'Boilermate' system using a combination unit.

Figure 3.7   Water-jacketed tube heater

3.4 Storage type water heaters and boiler heated systems

Domestic hot water supply installations of the storage type are either vented or unvented. Figures 3.8 and 3.9 illustrate the main features of each.

Vented hot water storage system

This is fed from a high-level cistern which provides the necessary pressure at outlets, accommodates expansion due to heated water, and is fitted with an open safety vent pipe to permit the escape of air or steam, and to prevent explosion without the need for any mechanical device.

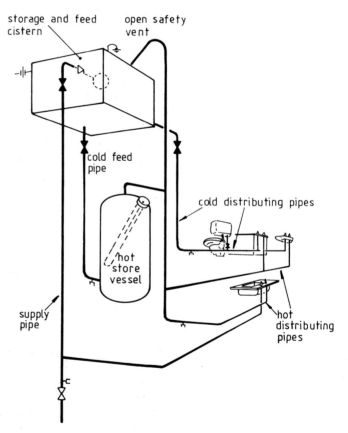

The open vent pipe will provide safety relief in the event of the system overheating.

Heated water will expand to the cold feed pipe.

The system shown is heated by immersion heater but could alternatively be heated by boiler.

Figure 3.8   Vented hot water system

The vented hot water storage system provides:

- constant low pressure;
- reserve water supply;

but needs:

- protection against the entry of contaminants to cistern (the 'protected' drinking water cistern to the requirements of Byelaw 30 will guard against this).

Unvented hot water storage system

This is usually fed direct from the supply pipe under mains pressure. It does not require the use of a feed cistern or open vent pipe, but relies on mechanical devices for the safe control of heat energy and hot water expansion.

Building Regulations require that hot water shall not exceed 100°C. The Approved Document recommends that all unvented hot water storage systems should be of the 'unit' or 'package' type, supplied complete with all safety devices, and should be fitted by an 'approved installer'.

A *unit system* (see figure 3.9(b)) has all safety devices and other operating devices fitted by the manufacturer at the factory, ready for site installation.

A *packaged system* (see figure 4.16) has all safety devices factory fitted. However, all other operating devices are supplied by the manufacturer in 'kit' form for site assembly.

Both the system and its installer should be 'approved' under the British Board of Agrément certification scheme.

Features of the unvented hot water storage system are as follows:

- eliminates the need for cold water storage and risk of frost damage;
- may require a larger supply pipe but eliminates some duplication of pipework;
- contains no reserve supply in case of supply failure;
- eliminates cistern refill noise;
- relies on mechanical controls which need regular inspection and maintenance;
- gives better pressure at outlets, particularly at showers;
- allows quicker installation than vented system but involves more costly components.

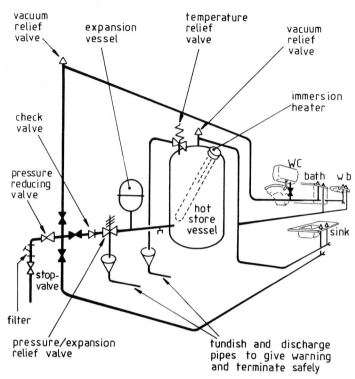

Pressure reducing valve reduces mains pressure to pressure suitable for operating system.

Pressure relief valve guards against excess pressure.

Expansion vessel accommodates expansion of water when heated.

Thermostat (not shown) controls temperature at normal level.

High temperature cut-out (not shown) protects against overheating of water.

Temperature relief valve allows boiling water and steam to escape if thermostat and thermal cut-out should fail.

Tundish and discharge pipes take relief water to safe place.

(a) Basic outline of system and components

Figure 3.9   Unvented hot water storage system

continued

Thermostat and high temperature cut-out fitted but not shown.

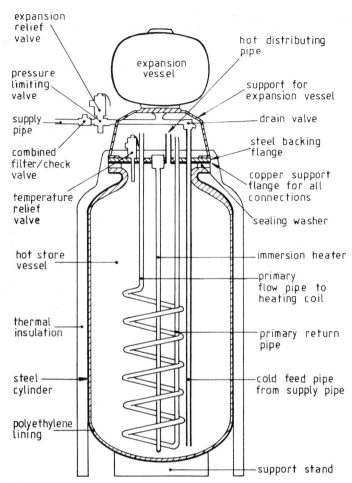

(b) Typical unvented hot water storage unit

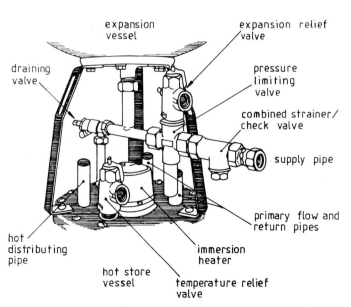

(c) Details of connections to unit heater

Figure 3.9   Unvented hot water storage system continued

Non-pressure or inlet-controlled water heaters

These are generally seen as single point heaters fitted either above the appliance with a swivel outlet spout, or under the appliance using special taps to control the flow before the heater inlet. They may be heated by either gas or electricity. See figure 3.10.

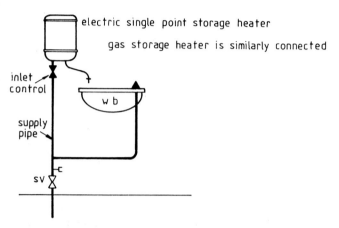

electric single point storage heater

gas storage heater is similarly connected

inlet control

w b

supply pipe

sv

Expansion of heated water overflows through outlet spout.

Outlet must not be obstructed, nor must any connection be made to it.

Can be connected to wash basin or sink if special taps are used to control the inlet and leave the outlet unobstructed.

Figure 3.10   **Non-pressure or inlet-controlled water heater**

Pressure or outlet-controlled water heaters

These may be heated by either gas or electricity. Although these are called pressure type heaters, they are generally designed for supply from a feed cistern and are not usually suitable for operation under direct mains pressure. See figures 3.11 and 3.12.

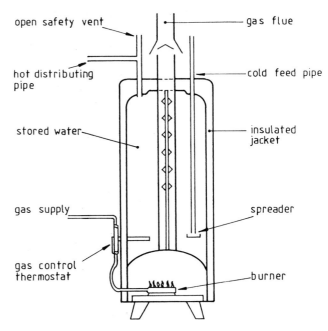

Figure 3.11   Outlet-controlled gas water heater

For dwellings, cistern must be 'protected'.

Heater capacity (minimum):
- for small dwellings, 100 l to 150 l,
- for off-peak electricity systems, 200 l.

Electricity systems are usually in the form of factory lagged and cased hot water cylinder with connections similar to those shown here for a gas installation.

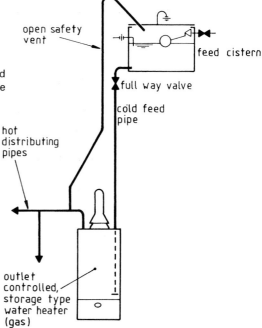

Figure 3.12   System using outlet-controlled heater

Cistern or combination type storage heaters

A combination type storage water heater incorporates a cold water cistern which should be located so that at the very least its base is not lower than the level of the highest connected hot water outlet, and is high enough to give adequate flows at outlets, see figures 3.13 and 3.14.

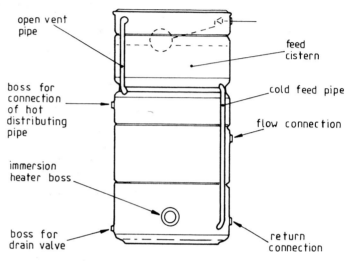

(a) Direct type

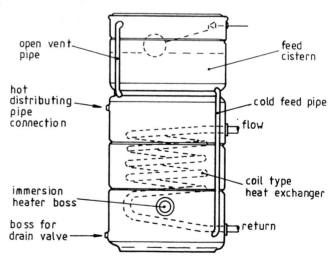

Also available:
  ○ factory lagged or purpose made unit with lagging and metal outer casing;
  ○ single feed indirect type.

(b) Double feed indirect type

Figure 3.13   **Combination storage heaters**

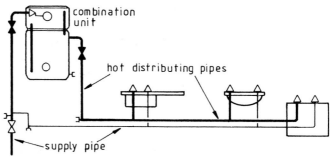

combination
unit

hot distributing pipes

supply pipe

Unit can be directly heated by immersion heater or indirectly heated from boiler.

Advantages:
- ○ low installation costs;
- ○ useful for flats when space is limited.

Disadvantages:
- ○ cold storage space limited;
- ○ low pressure at hot taps;
- ○ cannot be used for showers.

Figure 3.14   **Typical combination unit installation**

Electric immersion heater type storage heaters

Immersion heaters can be used as an independent heat source, or to provide supplementary heat to other centralised boiler systems, see figures 3.15–3.18.

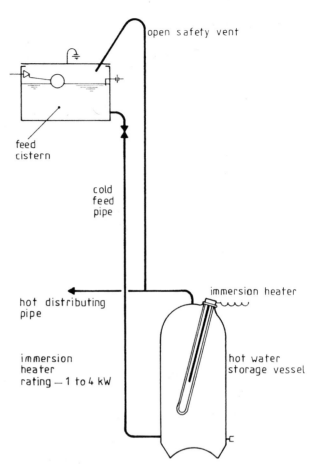

The hot store vessel may be a hot water cylinder or a combination unit, or a specially made, encased and lagged cylinder.

Figure 3.15   Storage system using immersion heater

(a) Single element – top entry

(b) Double element – top entry

Most common and cheapest arrangement. May not heat cylinder to bottom.

Thermostats and elements in hotter part of cylinder.

Dual provision from twin element to give full heat, or short boost from smaller element.

Switches and thermostat to comply with requirements of BS 3955.

(c) Single element – bottom entry

(d) Double element – twin side entry

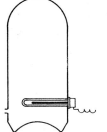

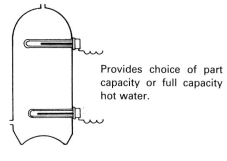

Better arrangement for supply of hot water with good temperature control.

Provides choice of part capacity or full capacity hot water.

Thermostats and elements in cooler part of cylinder.

Figure 3.16   Positioning immersion heaters

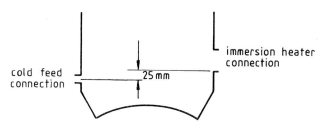

Immersion heater connection to be at least 25 mm above the centre line of the cold water inlet.

Adequate clearance needed around immersion heater for removal and replacement.

Figure 3.17   Immersion heater position and cold feed connection

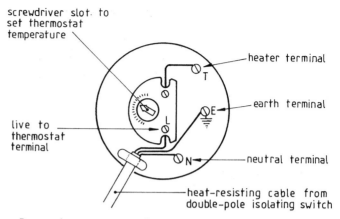

Range of temperature settings – 50°C to 82°C. Suggested setting 60°C.

Figure 3.18   **Top of immersion heater with cover removed**

Gas-fired circulators

These are essentially small gas-fired boilers. They may be independent or used in conjunction with a system using some other fuel, see figure 3.19.

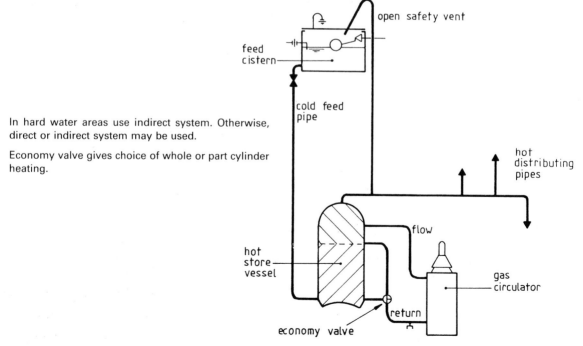

In hard water areas use indirect system. Otherwise, direct or indirect system may be used.

Economy valve gives choice of whole or part cylinder heating.

Figure 3.19   **System using gas-fired circulator**

Boiler-heated hot water systems

BS 6700 deals with independent water heating appliances using gas, oil, solid fuel and electricity, and includes direct, indirect, vented and unvented systems, see figures 3.20 and 3.21.

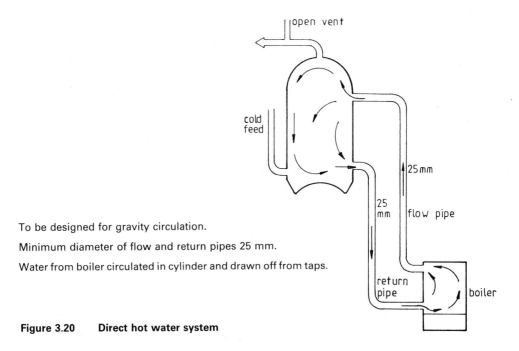

To be designed for gravity circulation.

Minimum diameter of flow and return pipes 25 mm.

Water from boiler circulated in cylinder and drawn off from taps.

Figure 3.20   Direct hot water system

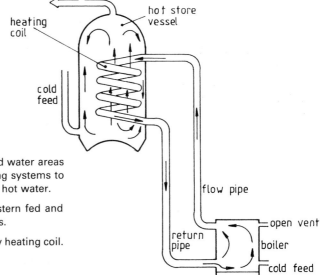

Indirect hot water systems should be used in hard water areas and for all combined hot water and central heating systems to avoid hard water scale and to maintain quality of hot water.

Primary circuits to indirect systems should be cistern fed and vented, or sealed using appropriate safety devices.

Water in boiler separated from water in cylinder by heating coil.

Figure 3.21   Indirect hot water system

Gravity flow hot water circulation

In gravity circuits good circulation depends on system height and pressure differentials between hot flow and cooler return, to create water movement and overcome frictional resistances in pipes and fittings, see figure 3.22.

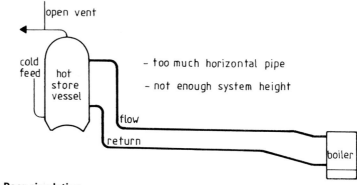

(a) Poor circulation

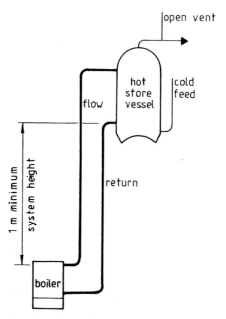

In gravity circuits good circulation depends on system height and pressure differentials between hot flow and cooler return to overcome resistances in pipes and bends.

(b) Good circulation

Figure 3.22   Circulation in primary circuits (gravity)

3.5   Primary circuits

Primary circuits are used for circulation of hot water between the boiler and the hot store vessel and include any radiator circuits (see figures 3.23 and 3.24). They may be vented or unvented. Circulation may be gravity or pumped.

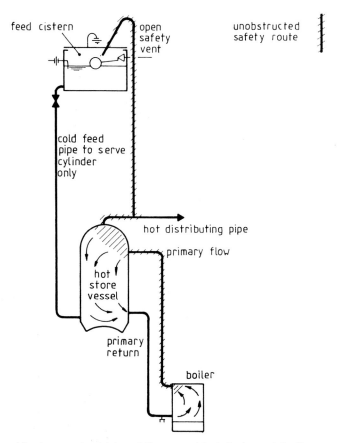

All pipes to be laid to falls to avoid air locks and facilitate draining.

Vent route from top of boiler through cylinder to open vent not to be valved or otherwise closed off.

Minimum sizes for primary circuits to hot store vessels:
- ○ 25 mm for solid fuel gravity systems;
- ○ 19 mm for small bore systems;
- ○ 13 mm for pump assisted systems.

Figure 3.23   Direct system of hot water (vented)

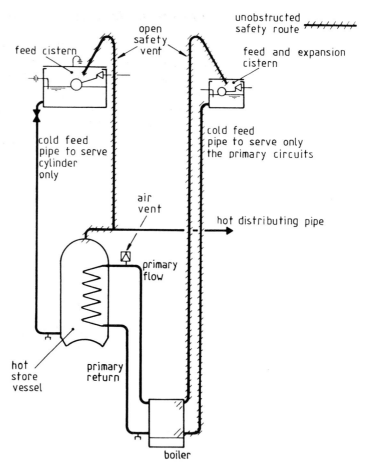

Open vent and cold feed pipes to primary circuit must *not* have valve, or be otherwise closed off.

The open vent and cold feed pipes may be connected to the primary flow and return pipes respectively.

Where vent pipe is not connected to highest point in primary circuit, an air release valve should be installed.

Figure 3.24   Indirect system of hot water (vented)

Hot water systems using the single feed indirect cylinder

The single feed indirect cylinder permits cost savings on materials and installation because there is no need for a separate open vent and cold feed for the primary circuits, see figures 3.25 and 3.26.

They are, however, limited in use as the volume of the cylinder is directly related to the volume of the water in the primary circuit (including radiators). If the system is oversized or becomes overheated they might lose their water seals (air locks) and revert to direct circulation. Single feed indirect cylinders must be vented.

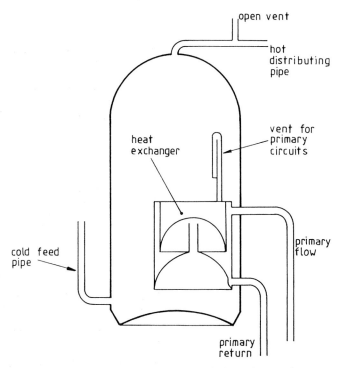

Expansion of heated water is taken up in inner heat exchanger, which must be matched by system size. Oversizing of system will result in loss of seals and mixing of primary and secondary waters.

Reference should be made to manufacturers regarding suitability for use with boilers and radiators.

Figure 3.25   **Single feed indirect cylinder to BS 1566**

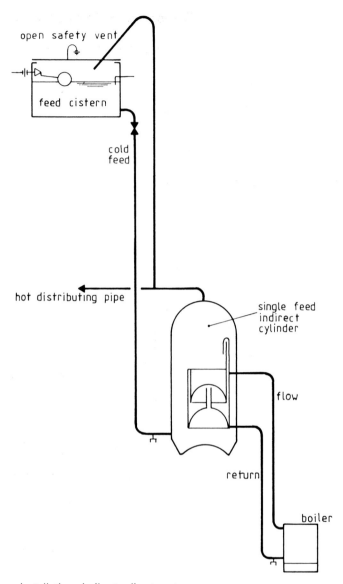

Installation similar to direct system.

When system is pumped, static head should exceed pump head.

No corrosion inhibitor to be added.

Figure 3.26   System using single feed indirect cylinder

Combined systems of hot water and central heating

These systems are shown in figures 3.27 and 3.28.

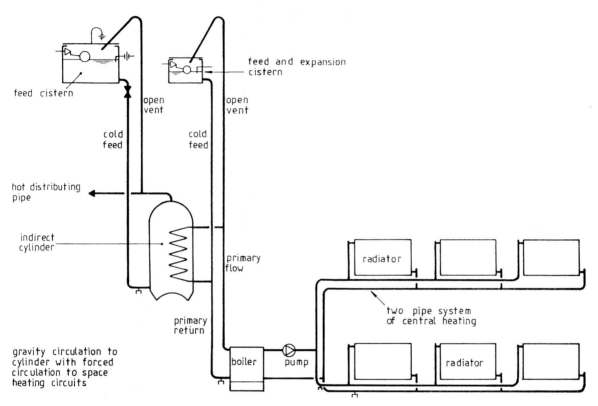

feed and expansion
cistern

feed cistern

open
vent

open
vent

cold
feed

cold
feed

hot distributing
pipe

indirect
cylinder

primary
flow

radiator

two pipe system
of central heating

primary
return

gravity circulation to
cylinder with forced
circulation to space
heating circuits

boiler

pump

radiator

Direct and uninterrupted route from boiler to feed and expansion cistern for cold feed and open vent pipes – also through
open vent from cylinder.

Figure 3.27   Combined hot water and central heating systems with vented primary circuits

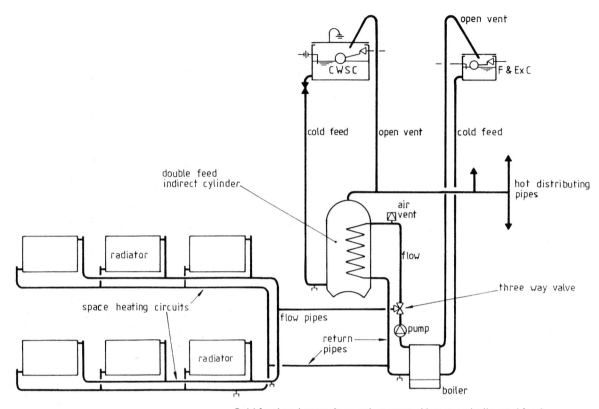

Cold feed and vent pipes uninterrupted between boiler and feed and expansion cistern.

Figure 3.28   **Fully pumped system of hot water and central heating (vented)**

Sealed central heating systems with vented hot water

This type of central heating system is shown in figure 3.29.

Indirect cylinder suitable for pressures of up to 0.35 bar in excess of relief valve setting.

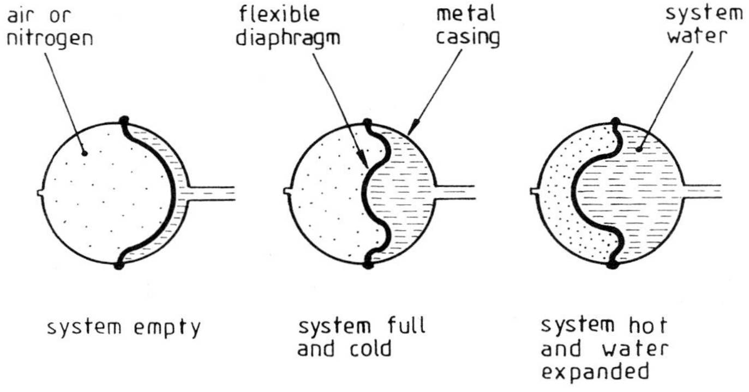

Sealed system may be used with gravity primaries to cylinder.

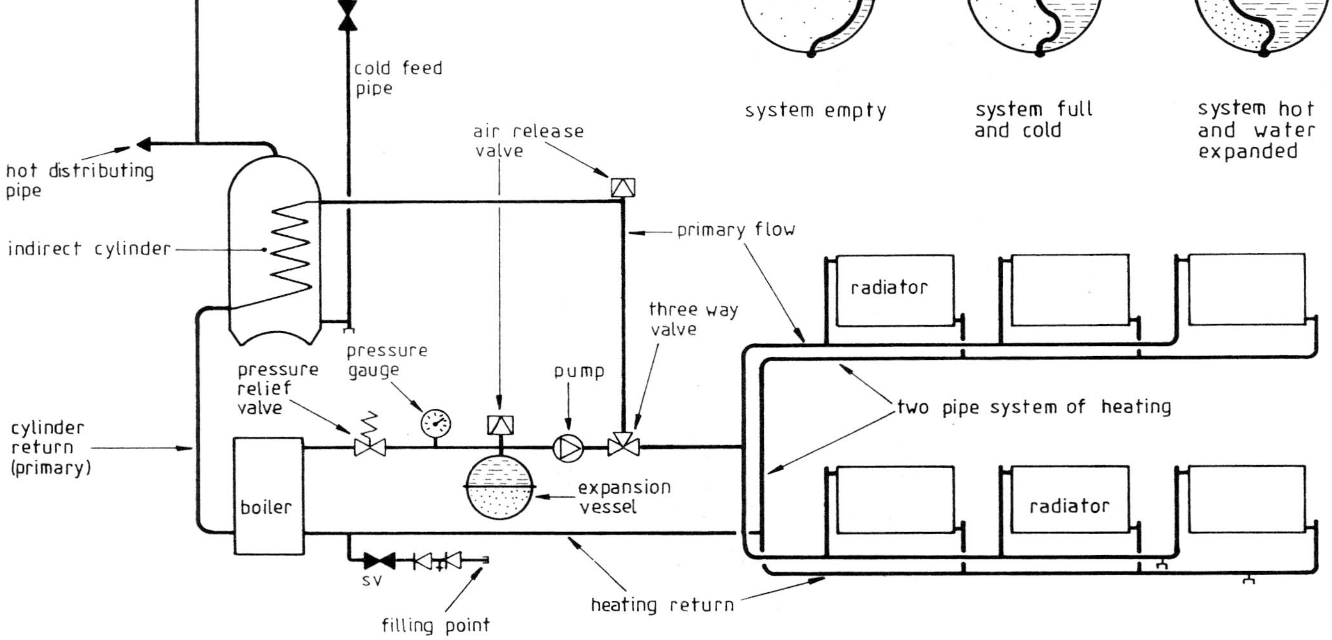

Detail from figure 3.29 to show operating principle of expansion vessel

Expansion vessel must accommodate expansion of the total volume of water in the whole system including space heating circuits (approximately 4%).

Figure 3.29 Sealed central heating system with vented hot water

Combined hot water and central heating with common return

This type of system is shown in figures 3.30 and 3.31.

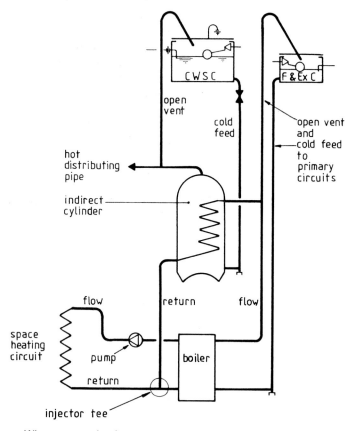

Where return circuits are connected via a common return to the boiler and only the heating circuit is pumped, an injector tee should be fitted.

Figure 3.30   **Combined hot water and central heating with common return**

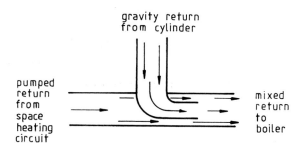

Figure 3.31   **Detail of injector tee**

Supplementary water heating and independent summer water heating

It is common practice for supplementary water heating and independent summer water heating to be provided by use of electric immersion heaters in the storage vessel or by gas circulators. Supplementary water heating provided from solar energy or heat pumps is growing in popularity.

Where supplementary electric immersion heating is to be used in conjunction with a boiler, the height of the storage vessel above the boiler should be sufficient to prevent circulation of hot water from the storage vessel to the boiler, when only the immersion heater is in use.

Solar heating

Solar energy heating can be useful to augment a conventional domestic hot water system (see figure 3.32) but, except on hot sunny days, cannot in Britain be considered sufficient to heat hot water to draw-off temperatures.

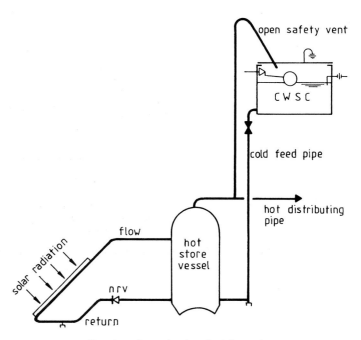

Drawing shows basic principles only.

Figure 3.32   **Solar heating system for hot water**

Many different designs of solar systems for heating water are possible, from simple direct feed gravity systems to more complex pumped circuits using two indirect storage cylinders or one cylinder with two indirect heating coils. Therefore, apart from indicating the possibilities of their use, they are beyond the scope of this book.

Heat pumps

Heat pumps extracting energy from the ground, water, or air at ambient temperatures, can be used to preheat conventional hot water systems, to augment existing systems, or to supply full hot water and central heating requirements, see figure 3.33.

Until such time as a British Standard is available giving recommendations for their use, guidance should be sought from heat pump manufacturers, or the Heat Pump Association, 161 Drury Lane, London WC2B 5QG.

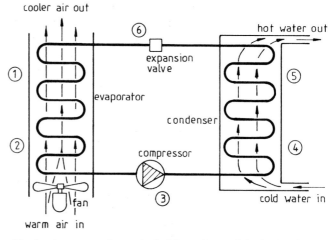

The heat pump works on the 'refrigeration cycle' principle (air to water)

(1) Refrigerant liquid forced under pressure into evaporator
(2) Heat energy from air absorbed by refrigerant as it is vaporized from liquid to gas
(3) Compressor 'squeezes' gas which becomes hotter and more compact
(4) The hotter compacted gas is pumped under pressure into the condenser
(5) In the condenser the gas becomes liquid and throws off heat which is absorbed into the hot water
(6) Condensed refrigerant liquid passes through expansion valve, to lower its pressure, before flowing through to evaporator to begin the cycle again

Figure 3.33   The heat pump – principle

3.6   Secondary distribution systems

A secondary hot water distribution system may be one of the following types:

(1) gravity fed (vented) from a cold water storage cistern;
(2) gravity fed from a water storage cistern;
(3) directly supplied under pressure from mains, through an instantaneous water heater or a water jacketed tube heater;
(4) unvented storage type, directly supplied under pressure from mains.

Connections to hot water storage vessels should be arranged so that the cold water feed pipe is connected near the bottom and hot water supply drawn off from the top of the cylinder, above any primary flow connection or heating element.

Dead legs and secondary circulation

To promote maximum economy of energy and water the hot water distribution system should be designed so that hot water appears quickly at draw-off taps when they are opened. The length of pipe measured from the tap to the water heater or hot water storage vessel should be as short as possible and should not exceed the lengths shown in figure 3.34. Where these lengths exceed the values given in table 3.1, the pipe should be insulated to at least the standard given in table 7.1. Insulation provided should ensure that heat losses from pipes do not exceed the values given in table 3.2.

Table 3.1   **Maximum lengths of uninsulated distributing pipes**

Outside diameter of distributing pipe mm	Maximum length m
Not exceeding 12	20
Exceeding 12 but not exceeding 22	12
Exceeding 22 but not exceeding 28	8
Over 28	3

Table 3.2   **Maximum permitted rates of heat loss from pipes**

Outside diameter of pipe* mm	Maximum heat loss W/m^2
10	675
20	400
30	280
40	220
50 and above	175

* For intermediate values of pipe diameter, the corresponding maximum heat loss is found by linear interpolation.

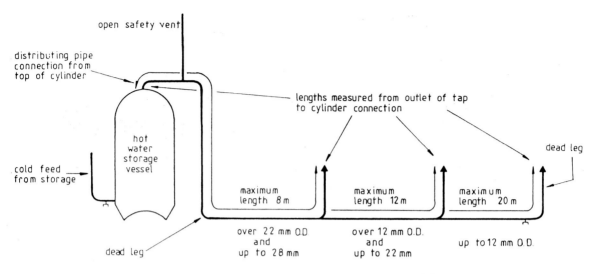

O.D. means outside diameter

Where lengths of dead legs exceed those shown, pipes should be insulated.

Note. 'Dead leg' means the length of distributing pipe without secondary circulation.

Figure 3.34   Maximum length of dead legs without insulation

Secondary circulation minimises delays in obtaining hot water from taps and reduces waste of water during any delay. Secondary circulation should be considered when short dead legs are impractical and the circuit should be well insulated to reduce the inevitable heat losses from pipe runs. A diagrammatic arrangement of secondary circulation is shown in figure 3.35.

In systems where it is not possible to attain gravity circulation, a non-corroding circulating pump should be installed to ensure that water within the secondary circuit remains hot. The pump should be located on the return pipe close to the cylinder.

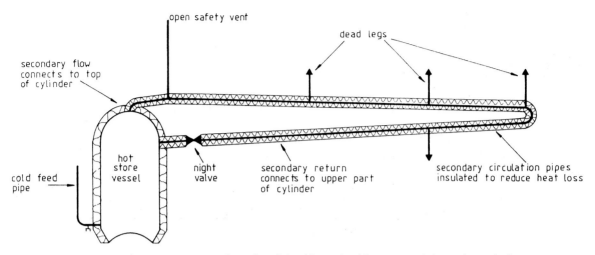

open safety vent

dead legs

secondary flow connects to top of cylinder

hot store vessel

cold feed pipe

night valve

secondary return connects to upper part of cylinder

secondary circulation pipes insulated to reduce heat loss

Lengths of dead legs should not exceed those shown in figure 3.34.

Dead legs feeding spray taps should not be more than 1 m long.

Figure 3.35   **Distribution system with secondary circulation**

3.7   Components

Cold feed pipes

A cold feed pipe supplies hot water apparatus with cold water from storage (see figure 3.36). In vented systems of secondary hot water, the cold feed water is supplied from a feed cistern.

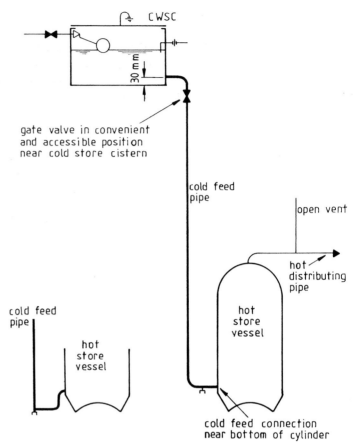

gate valve in convenient
and accessible position
near cold store cistern

cold feed
pipe

open vent

hot
distributing
pipe

cold feed
pipe

hot
store
vessel

hot
store
vessel

cold feed connection
near bottom of cylinder

The dipped entry connection reduces the risk of hot water
circulation in the cold feed pipe.

Cold feed must serve hot water apparatus only and no other
appliance.

The cold feed pipe must *not* be connected to the boiler on direct
systems.

No valve to be fitted to cold feed pipe on primary circuits.

Figure 3.36   Cold feed pipes

In a vented indirect system, other than one using a single feed indirect
hot water cylinder, the primary circuit must be supplied through a
separate cold feed pipe from a separate feed and expansion cistern. The
cold feed pipe should be connected to the lowest point in the primary
circuit near to the boiler.

The open vent pipe

The open safety vent, as it is often called, is connected to the top of the hot water storage vessel and rises to terminate above the cold feed cistern. Recommendations for the installation of the open safety vent are given in figure 3.37.

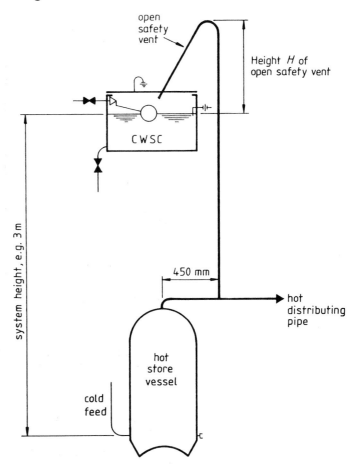

Vent connection offset to reduce circulation and loss of heat in vent pipe.

Rules for installation of the open vent:
- No valve to be fitted to vent pipe.
- Pipe to rise continuously from connection at hot store vessel to highest point over cistern.
- Vent pipe not less than 19 mm bore.
- Vent pipe *not* to be connected to cold feed.
- Formula for height H of open safety vent over cistern to prevent expanded water from overflowing:

$H = 150$ mm $+ 40$ mm per metre of system height

For a system height of 3 m

$H = 150 + (3 \times 40)$

$= 270$ mm

Figure 3.37   The open safety vent

Vented primary circuits are similarly treated. The vent must terminate over a feed and expansion cistern. Due allowance should be made for any head produced by a circulation pump (if fitted) in order to prevent either the continuous discharge from the vent, or air entrainment into the system.

The lower connection of the vent may be made to the highest point in the primary flow pipe, or preferably to the top of the boiler.

Cylinders supplied from common feed cisterns

Common cold feed pipes are permissible, provided the installation is as shown in figure 3.38 and where individual separate systems are not practicable.

Separate vents are required to prevent hot water circulation from one cylinder to another.

Anti-siphon pipe is shown as an alternative backflow prevention device and is not needed if mechanical devices are used at each floor level.

Pipes to be aligned so as to avoid air locks and facilitate filling and draining.

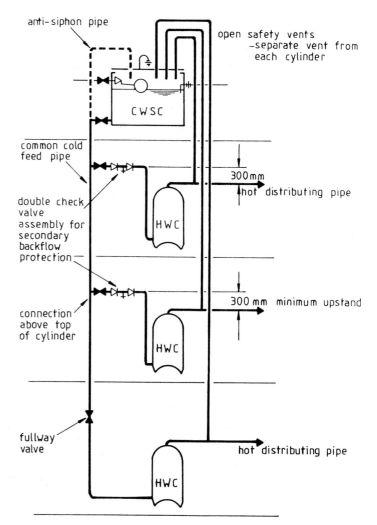

Figure 3.38   Common cold feed arrangement

Hot water storage vessels

Hot water storage vessels (see figure 3.39) are manufactured in a number of varieties and styles. Except in special cases, only cylinders and tanks to a relevant British Standard should be considered, preferably one with a works-applied external insulation.

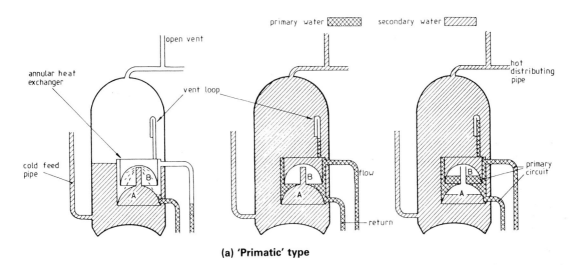

(a) 'Primatic' type

System filling Water from cold feed enters cylinder and from there into inner tank A. It then passes through inner tank B before finally filling the annular heat exchanger and the primary circuits. Air escapes from the vent loop.

System full Inner tank A is full. Inner tank B retains an air pocket. Annular heat exchanger is full and vent loop retains an air pocket. The air pockets prevent mixing of primary and secondary waters.

System heating up Hot water expands in primary circuit and annular heat exchanger. The expanded water pushes the air pocket from tank B to be retained in tank A and the air pocket in the vent loop is moved but also retained.

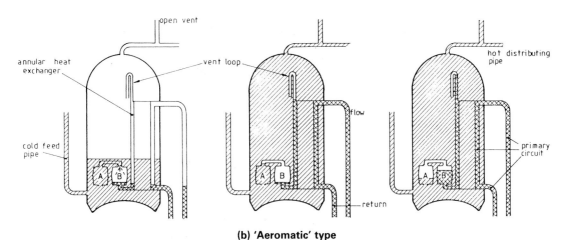

(b) 'Aeromatic' type

Figure 3.39   **Single feed indirect cylinders – working principles**

Hot water storage vessels should be arranged so that the hotter water at the top floats on the relatively cold water, and hot water can be drawn off even though a substantial quantity of cold feed water may have recently flowed into the vessel.

Cylinders and tanks should be installed with the long side vertical to assist effective stratification or 'layering' of hot and cold water (see figure 3.40). The ratio of height to width or diameter should not be less than 2:1. An inlet baffle should be fitted, preferably near the cold inflow pipe, to spread the incoming cold water.

Hot water storage capacities should be related to likely consumption and the recovery rate of hot water storage vessel.

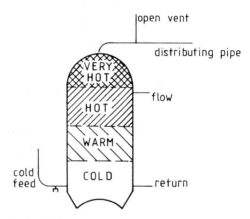

(a) Vertical cylinder

Vertical cylinder preferred to give better stratification and ensure that the hottest water is drawn off.

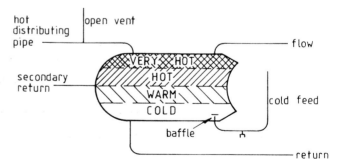

(b) Horizontal cylinder

Baffle will help prevent water from cold feed mixing with hot water at top of cylinder.

Hot water layer is very shallow and when drawn off may allow water layers to mix.

Figure 3.40   Cylinders showing stratification

For domestic use the temperature of stored hot water should not exceed 65°C and must not reach 100°C at any time. A temperature of 60°C should be adequate to meet all normal domestic requirements.

In other cases, such as in larger kitchens and in laundries, water temperatures may need to be higher. However high temperatures should be avoided wherever possible because of furring in hard water areas and the danger of scalding users. See table 3.3 for domestic hot water requirements.

Table 3.3   **Hot water supply requirements**

Requirement (dwellings)	Quantity l	Temperature °C	Flow rate l/s
Stored water per dwelling	135	60	—
Hot water used per person per day	35 to 45	Various	—
Bath per use	100	40	0.03 per tap
Shower per use	25	32 to 40	0.01 mixed
Wash basin per use	4.5	40 to 60	0.15 per tap
Sink per use	18	60	0.2 per tap

Feed and expansion cisterns (see figure 3.41) should comply with the requirements of BS 417 and BS 4213 and with the byelaw requirements for cold water storage cisterns.

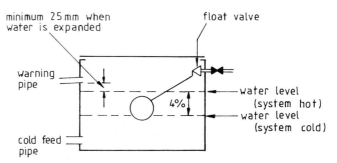

Space must be allowed to accommodate expansion equal to 4% of water in circuit

Cistern and float valve to resist temperature of 100°C

Figure 3.41   **Feed and expansion cistern**

Boilers

All boilers, firing equipment, and components of a combined boiler and cylinder unit should comply with appropriate British Standards wherever possible.

General provisions relating to boilers are as follows:

(1) Domestic boilers should be powerful enough to heat stored water to 65°C in $2\frac{1}{2}$ hours or less, including any towel rails, airing coils and secondary circulating pipes.

(2) Boilers and rooms containing boilers must be adequately ventilated to provide:
  (i) sufficient air for combustion;
  (ii) enough air to remove the products of combustion safely and properly without danger to occupants of the building (see Building Regulations and the Gas Safety Regulations).

(3) Adequate space for access (see figure 3.42) is needed for:
  (i) boiler maintenance, removal of burners, pumps and pipe connections, and its eventual replacement;
  (ii) stoking and cleaning, i.e. 1.25 times the back to front dimension of the boiler, subject to a minimum of 1 m.

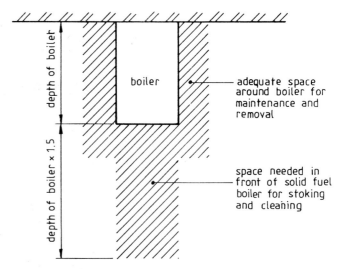

Figure 3.42   Space requirements for access to boilers

(4) Precautions should be taken to ensure that combustible construction materials are not placed near boilers.

(5) Fuel stores should be at a safe distance from the boiler.

(6) Precautions should be taken to prevent explosion should water temperatures exceed 100°C. An explanation of these precautions follows in chapter 4.

(7) In all cases, boiler manufacturers' recommendations should be followed for correct installation, commissioning and maintenance.

(8) Solid fuel boilers produce heat energy for a long time after being stoked and cannot readily be shut down. Sufficient allowance should be made for the safe dissipation of that heat within the hot water or heating system.

Boilers for use with sealed primaries and unvented systems require specific controls and must therefore be selected only from those suitable specifically for these purposes.

Gas-fired boilers should be appropriate for the gases with which they are to be used.

Oil-fired boilers should be selected from the Domestic Oil Burning Equipment Testing Association (DOBETA) list of tested and approved domestic oil-burning appliances.

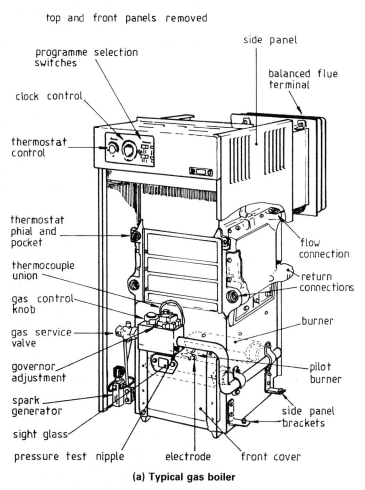

(a) Typical gas boiler

Figure 3.43   Boilers

continued

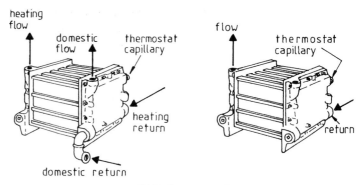

(b) Boiler connections

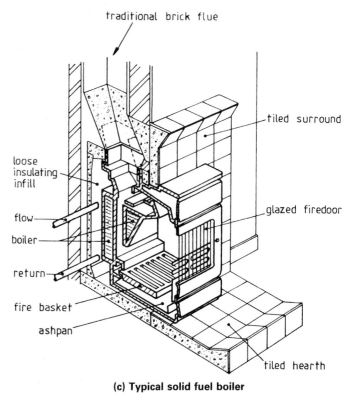

(c) Typical solid fuel boiler

Figure 3.43   **Boilers**
continued

Flues for boilers and water heaters

BS 6700 does not give details of requirements for flues but refers to various other codes and in particular to the Gas Safety Regulations and to the Building Regulations. In general terms the standard makes the following three points.

(1) An adequate flue shall be provided wherever necessary.

(2) All materials and components shall comply with the requirements of the appropriate standard.

(3) The use of an inadequate or badly maintained flue can have fatal consequences.

Circulating pumps

Pump circulation is needed in all cases where natural circulating pressure is insufficient (see figure 3.44.)

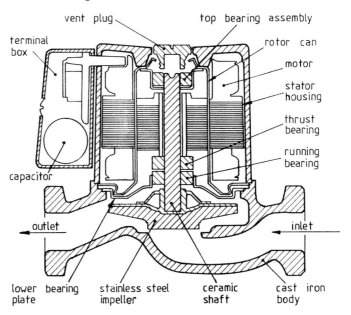

(a) Section through circulating pump

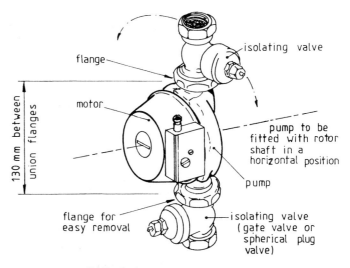

(b) Typical pump arrangement

Figure 3.44   **Circulating pump**

Immersed rotor (glandless) type circulating pumps must be used on primary circuits only. Pumps used for secondary circulation must be resistant to corrosion.

Inlet and outlet connections should be fitted with fullway valves, and space allowed for renewal or repair.

Circulating pumps should be quiet in operation and suitably suppressed to prevent radio or television interference.

An electrical isolating switch must be fixed adjacent to and within sight of the circulating pump. The whole of the wiring, earthing, etc. must be carried out in accordance with the current Regulations for Electrical Installations of the Institution of Electrical Engineers.

Circulating pumps should comply with the requirements of BS 1394: Parts 1 and 2, and should be installed in accordance with manufacturers' recommendations.

Valves and taps

Valves used for isolating a section of the water service should provide a positive seal when closed. See also servicing valves, in section 2.4.

Pressure operated, temperature operated and combined relief valves, check valves, pressure reducing valves, anti-vacuum valves and pipe interrupters should be fitted in accordance with Building Regulations and water byelaws (see chapters 4 and 5).

Draining taps should comply with the requirements of BS 1010 or BS 2879 (see figure 7.9), and be suitable for hosepipe connection.

Although water byelaws require draining taps only on the supply pipe, it is recommended that they should also be fitted so that the entire water system can be drained down when not in use, thus avoiding frost damage.

A further advantage is that systems can be readily drained before repair work is carried out so reducing the risk of mess and damage from pipes and fittings that are disconnected (see also chapter 7).

Mixing valves

Valves used for the mixing of hot and cold water (see figures 3.45 to 3.47) must be connected so that protection is provided against the possible contamination of water supplies.

Tap mixers or combination taps

There are two types of tap mixers:

(1) Those that mix water within the valve body. These should be supplied from the same source, i.e. both hot and cold from a common storage cistern, or both under direct mains pressure. This ensures that no crossflow can occur between water sources, and gives better and safer temperature and flow control.

(2) Twin outlet types that mix water outside the valve outlet. These are of two designs. First, the divided outlet which creates no crossflow risk and can be connected to differing sources, and second, the type with a tube within a tube which is suitable for supply from separate sources, provided care is taken to ensure that the cold is on the outside, and the hot on the inside (see figure 3.46). This also reduces the risk of scalding.

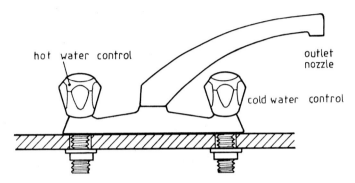

Figure 3.45   **Sink mixer tap**

sectional view through outlet nozzle

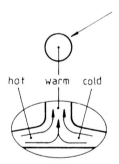 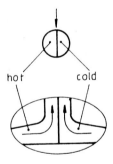 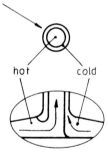

Internal mixing type

Mixing occurs in body of valve.

Precautions required against crossflow.

Difficult to maintain temperature where hot and cold inlet pressures vary.

Divided outlet with external mixing

Prevents crossflow between hot and cold supplies.

Some risk of scalding from hot outflow.

Double tube arrangement with external mixing

Prevents crossflow provided cold is on the outside and hot on the inside.

Reduced risk of scalding from outflow.

Figure 3.46   **Types of mixer tap**

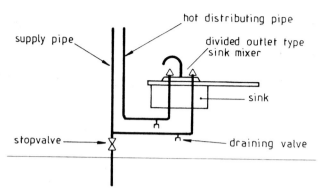

Hot and cold supplies from mixed sources may be connected as shown if the mixer is of the divided outlet type.

Where water is blended within the mixer, supplies should be from a common source. However, the above arrangement may be permitted if the hot water is supplied from a cistern complying with Byelaw 30.

Figure 3.47   **Sink mixer installation**

Shower mixers

There are two types: thermostatic and non-thermostatic. Thermostatic types (see figure 3.48) are preferred because they provide automatic temperature control. They sense changes in temperature and adjust hot and cold flows accordingly. They also provide the user with a degree of protection against the effects of irregular flow and pressure variations caused when other taps are used.

Thermoscopic (thermostatic) assembly provides temperature balance.

Shuttle and sleeve assembly will shut off hot water in the event of cold water failure.

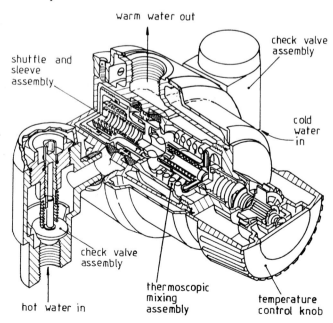

Figure 3.48   **Thermostatic shower mixer**

(a) Storage-fed shower

Separate supplies to shower:
- decrease risk of flow and pressure variations;
- eliminate backflow risk.

Shower mixer supplied from one source, e.g. hot and cold from storage or both hot and cold direct from mains.

With a fixed shower head which has an air gap maintained at all times, there is no backflow risk.

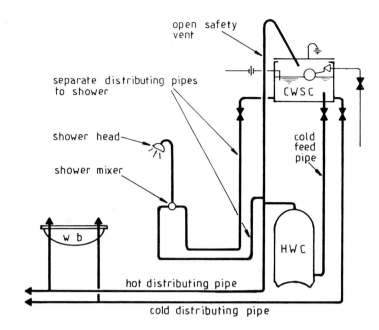

(b) Mains fed shower

Where shower hose is constrained and type A air gap is maintained at nozzle at all times, backflow prevention devices such as the double check valve assembly are not needed.

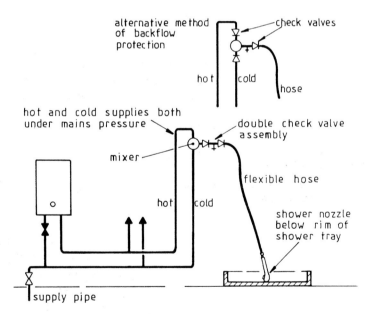

Figure 3.49   **Shower installation**

3.8   Energy supply installations

The wiring to electric heaters must be in accordance with the current edition of the Regulations for Electrical Installations published by the Institution of Electrical Engineers, and installed by a person such as a certificate holder of the National Inspection Council for Electrical Installation Contracting (NICEIC).

All gas installation work must comply with the Gas Safety (Installation and Use) Regulations 1972 and be carried out in accordance with relevant Codes of Practice. Installers must be registered with the Council for the Registration of Gas Installers (CORGI) and approved for the category of work concerned. Any person working on gas must be competent in gas work.

Solid fuel installations must be installed in accordance with the relevant Codes of Practice by a person such as a Solid Fuel Advisory Service registered heating contractor.

Chapter 4
Prevention of bursting

It is extremely dangerous to heat water in a filled enclosed vessel to a temperature above 100°C. Water boils at 100°C at normal atmospheric pressure, but at 3 bar pressure, for example, the boiling point rises to 143°C and the heated water will have expanded and/or its pressure will have increased. If the water expands to such an extent that a small split develops in the cylinder, there will be an immediate drop in pressure and the water will flash to steam increasing in volume by about 1600 times. This will result in the total rupture or explosion of the vessel.

Building Regulations require that the temperature of stored hot water shall not exceed 100°C.

Successful and safe operation of a hot water system depends on the following:

- the right equipment, properly installed and maintained, and not exposed to misguided interference;
- reliability and durability of safety devices and equipment;
- location and choice of system and components;
- Kitemarked equipment, or equipment approved by the Water Research Centre, or, where appropriate, systems and safety devices approved under the British Board of Agrément scheme for unvented systems;
- good maintenance, for continued safety, with reasonable expectation that this will continue;
- good control of temperature and pressure;
- efficient control and use of energy.

4.1 Energy control

Three forms of energy control are accepted by the Building Regulations approved documents: thermostatic control; temperature relief; and heat dissipation.

Thermostatic control

Effective thermostatic control is needed to prevent the temperature of stored water from rising above the normal expected hot water temperature of 60°C to 65°C. This may be achieved by using either a cylinder thermostat (see figure 4.1) or an immersion heater thermostat (see figure 4.2). Thermostatic control is used for the normal control of heat in hot water storage vessels, boilers and heating appliances (see figures 4.3 and 4.4.)

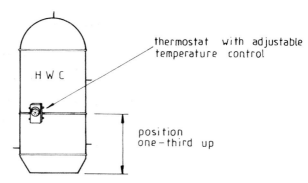

Figure 4.1   Cylinder thermostat

The permanent magnet ensures positive making and breaking of the contacts. This avoids arcing with consequent burned contacts and possible fire risk.

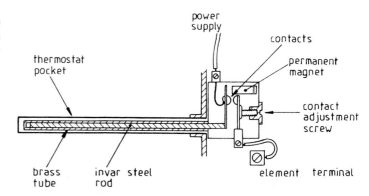

Figure 4.2   Section through immersion heater thermostat

Heat cannot be shut off immediately as fuel takes time to cool down.

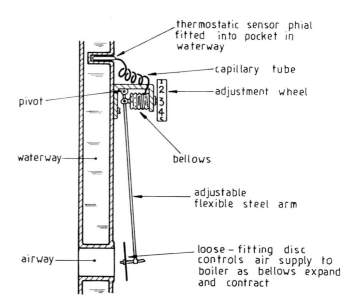

Figure 4.3   Thermostatic control for small solid fuel boiler

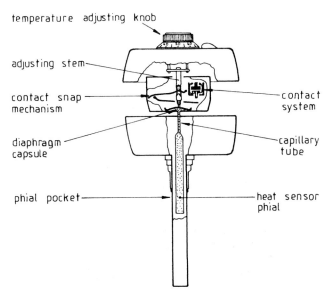

temperature adjusting knob

adjusting stem

contact snap
mechanism

contact
system

diaphragm
capsule

capillary
tube

phial pocket

heat sensor
phial

Used in gas or oil burners to give instant control.

Figure 4.4   Boiler thermostat, liquid filled

Temperature-operated energy cut-out device

This device is designed to cut off the heat source at a predetermined maximum temperature of 90°C. It must be of the non-self resetting type, and will only operate if the normal temperature thermostat fails and the hot store vessel overheats, see figure 4.5.

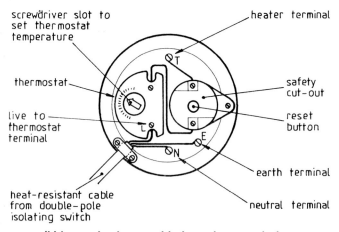

reset button

triple pole
contact system
(circuit breaker)

spring loaded
snap arm

bi-metallic
element

heat sensor
phial

phial pocket

For use with oil or gas burners.

(a) Boiler cut-out device

screwdriver slot to
set thermostat
temperature

heater terminal

thermostat

safety
cut-out

live to
thermostat
terminal

reset
button

earth terminal

heat-resistant cable
from double-pole
isolating switch

neutral terminal

(b) Immersion heater with thermal cut-out devices

Figure 4.5   High energy cut-out devices

Temperature relief and heat dissipation

Adequate means of dissipating the heat input are needed in case both
the temperature thermostat and the energy cut-out fail. Two methods
are accepted by Building Regulations.

(1) A **vent pipe** to atmosphere, often known as the open safety vent,
    seen as part of the traditional vented hot water system, and shown
    in figures 3.23 and 3.24.

(2) A **temperature relief valve** (see figure 4.6) as used for many years in unvented systems throughout the world. This must be fitted in the top of the storage vessel, within the top 20% volume of the water. This device is used as a safety back-up in case both the temperature thermostat and the thermal cut-out device fail.

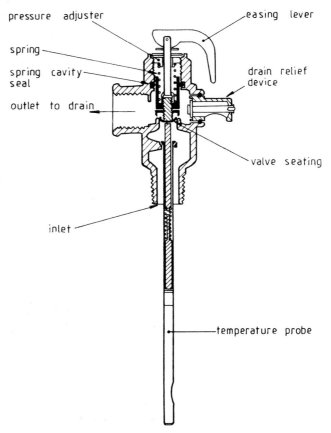

Discharge capacity 1.5 times that of maximum energy input to heater.

No valve to be fitted between temperature relief valve and heater.

Figure 4.6   **Temperature relief valve to BS 6283: Part 2**

The water discharged from a temperature relief valve must be removed from the point of discharge to a safe place. In operation the valve will open at a pre-set temperature to permit the overheated water to escape safely from the hot water storage vessel before it boils. It will usually operate at about 95°C.

Maintenance and periodic easing of temperature relief valves is particularly important for continued efficiency. A notice drawing attention to this should be provided in a prominent position for the user.

Sealed primary circuits

When sealed primaries are used; it is permissible for a second temperature-operated energy cut-out to be employed in place of the temperature relief valve. Sealed primary circuits must not be heated by solid fuel. Both a vent and a temperature relief valve must be fitted in addition to a thermostat if water in a primary circuit or in a direct system is heated by solid fuel. This is because complete thermostatic control and an effective temperature-operated energy cut-out are not available.

Discharges from temperature relief valves and expansion/pressure relief valves

Building Regulations state 'there shall be adequate precautions to ensure that the hot water discharged from safety devices is safely conveyed to where it is visible but will cause no danger to persons in or about the building'.

Two important points are stated here.

(1) Any discharge from a temperature relief valve or expansion relief valve must be readily visible. This will:
    (i) show there is a fault on the system that requires maintenance;
    (ii) reduce wastage of water because faults will be seen and rectified.
(2) Any discharge must be to a safe place. This will apply more importantly to the temperature relief valve, which will discharge very hot water from the top of the hot storage vessel, whereas the discharge from the pressure relief valve situated on the cold supply pipe will be relatively cold. See figures 4.7 and 4.8.

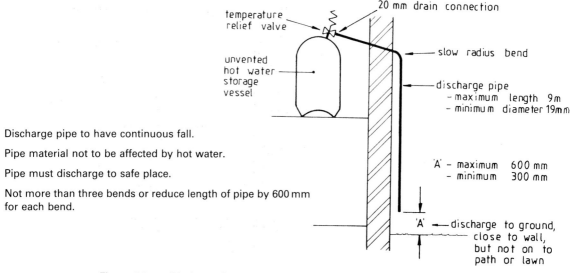

Figure 4.7   Discharge from temperature relief valve

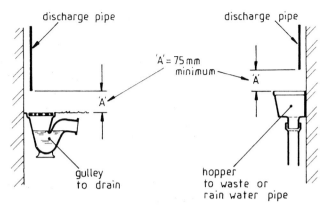

Consider suitability of drain or down pipe material for hot water.

(a) To gulley   **(b) To RWP hopper**

Figure 4.8   Alternative methods of discharge

The use of a tundish (see figure 4.9) will permit greater flexibility in the positioning of appliances and discharge outlet and will also provide visibility where discharges are more than 9 m from the appliance.

Discharges from expansion or pressure relief valves can be treated a little more leniently than those from temperature relief valves as the danger from the discharge is considerably less.

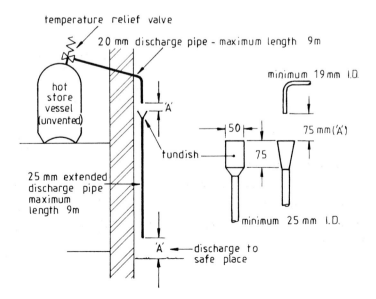

Figure 4.9   Positioning and design of tundish

Additional considerations for multiple installations, e.g. flats

The problems here are ensuring that visibility is maintained and also knowing which appliance is at fault should a discharge occur, (see figure 4.10).

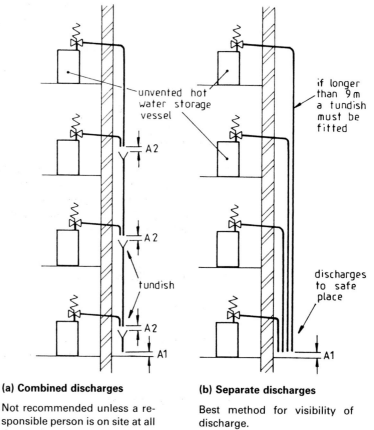

(a) Combined discharges

Not recommended unless a responsible person is on site at all times.

(b) Separate discharges

Best method for visibility of discharge.
A1 = 150 mm minimum if to ground
A2 = 75 mm minimum

Figure 4.10   Discharges from multiple installations

Pipework may be within the building, provided that tundishes and discharge terminals are readily visible and adequate provision is made to remove any discharges from the building safely, e.g. by trapped connections to drains, see figure 4.11.

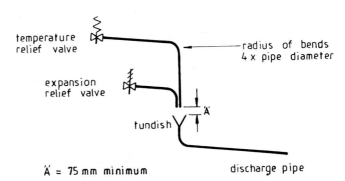

'A' = 75 mm minimum   discharge pipe

Discharge to safe and visible position.

Figure 4.11 **Combined discharge from temperature relief valve and expansion relief valve**

Sequence of operation

Thermostats, temperature-operated cut-outs and temperature relief valves must be set to operate in this sequence as the temperature increases. These three devices are not essential where water is only heated indirectly by a primary circuit which is already protected, or from a source of heat that is incapable of raising the temperature above 90°C.

4.2 Pressure and expansion control

Working pressures

In any system the working pressure must not exceed the safe working pressure of the component parts, (see table 4.1).

Table 4.1 **Safe working pressures**

Circuit	Maximum working pressure bar	Test pressure bar
Sealed primary	3	5
Unvented secondary	6	10

Where necessary the supply pressure should be controlled by using a break cistern or pressure reducing valve, (see figures 4.12 and 4.13).

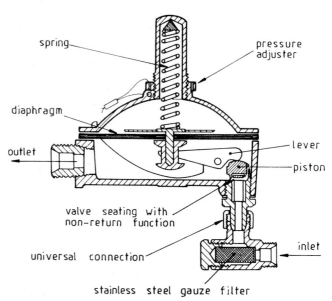

Gives finite control, and constant outlet pressure allowing accurate sizing of system.

Essential where copper hot water cylinder is used.

Figure 4.12   Diaphragm-actuated pressure reducing valve, to BS 6283: Part 4

Provides cruder control than the diaphragm-actuated reducing valve, to maintain supply line at preset maximum pressure.

Use only with glass lined steel (or similar) cylinder on higher pressure system (up to 6 bar).

Not for use with copper cylinder on low pressure system (up to 3 bar).

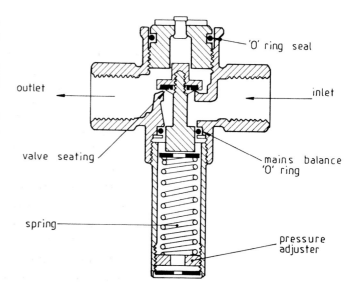

Figure 4.13   Pressure limiting valve

Control of water expansion

Within any hot water system the expansion of the heated water and consequent possible pressure rise must be limited without any discharge of water to waste. That is, expansion water must be accommodated within the hot water system. There are a number of ways of doing this depending upon the system (vented or unvented) and the size of the water heater and/or any storage vessel.

(1) In the traditional vented system expanded water is permitted to move back into the cold feed pipe towards the feed cistern (see figure 4.14) but only as long as the cold feed pipe is unobstructed so that there will be no resistance to the flow of expanded water. Consequently, there will be no increase in pressure in the supply due to expansion.

(2) In the unvented hot water storage system expansion control is normally achieved by using an expansion vessel which should be

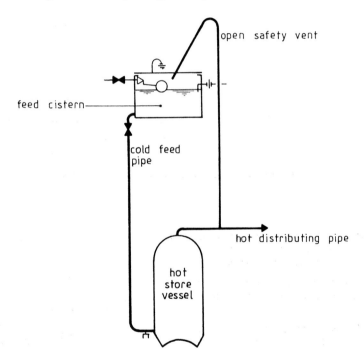

Feed cistern also serves as break cistern to limit pressure in system.

Valve not to have loose jumper.

Cold feed not to be fitted with check valve or other obstruction.

Open safety vent to permit escape of steam in the event of system overheating.

Vent to be open at all times.

Heated water expands safely into cold feed pipe.

Figure 4.14   Expansion in a traditional vented system

rated to accommodate a volume of expanded water at least equal to 4% of the total volume of water likely to be heated in the system. (See figures 4.15 and 4.16). The expansion vessel should be connected on the cold water inlet to the storage vessel.

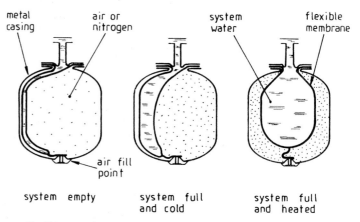

system empty   system full   system full
                      and cold   and heated

Flexible membrane prevents contact between water and steel casing to minimise corrosion.

Figure 4.15   Expansion vessel

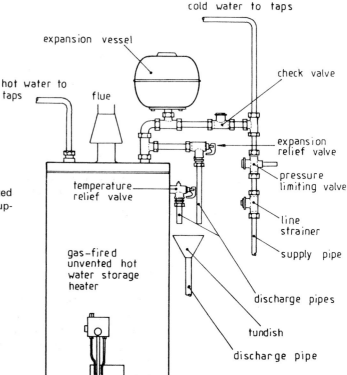

Thermostat and thermal cut-out factory fitted but not shown. Other devices shown are supplied with package for site assembly.

Figure 4.16   Use of expansion vessel in packaged unvented hot water system

(3) Reversed flow into supply pipe. In any mains supplied unvented hot water supply system whether instantaneous, water-jacketed tube or storage type, reverse flow along the supply pipe may be permitted (see figure 4.17), provided that there is no restriction on the supply pipe such as a check valve, pressure reducing valve or stopvalve which might prevent the reversal of flow as and when water expands. The supply pipe should be large enough to accommodate a volume of water at least equal to 4% of the total volume of water to be heated. In these circumstances, no heated or warm water should reach either the communication pipe or any branch pipe feeding a cold water outlet.

This method, although permitted by water byelaws, is not widely practised, particularly in systems installed in older properties where the undertaker's stopvalve might incorporate a loose jumper.

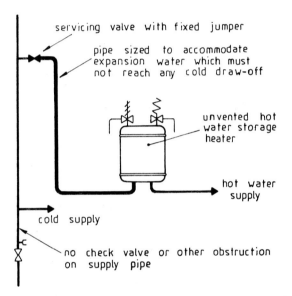

servicing valve with fixed jumper

pipe sized to accommodate expansion water which must not reach any cold draw-off

unvented hot water storage heater

hot water supply

cold supply

no check valve or other obstruction on supply pipe

Not recommended because of the possibility that an obstruction to supply pipe could be fitted at a later date.

Figure 4.17   System using supply pipe for expansion

(4) An expansion relief valve (see figure 4.18) may be used as a fail safe device. However, it is not permitted to be used as the sole means of control of expansion water. This valve should be arranged to open automatically and to discharge water only when the pressure in the system reaches a predetermined level under failure conditions.

Any water discharged from an expansion valve must be discharged safely to a conspicuous position in a similar fashion to that of the temperature relief valve (see section 4.1).

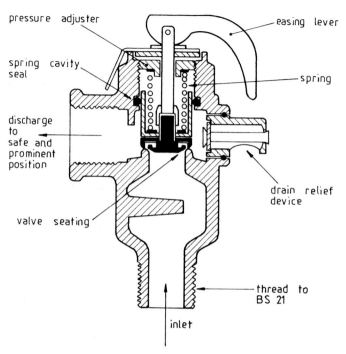

pressure adjuster

easing lever

spring cavity seal

spring

discharge to safe and prominent position

drain relief device

valve seating

thread to BS 21

inlet

Figure 4.18   **Expansion relief valve**

4.3   Control of water level

Connections should be made so that water cannot be drawn (1) from any primary or closed circuit, and (2) to ensure that the hottest water can be drawn only from the top of a hot water storage vessel and above any primary flow connection or heating element. (See figure 4.19.) This includes the position of connections from secondary returns.

Unintentional draining down of any system (particularly storage types) is dangerous and should be avoided, as it may:

- expose temperature sensing controls, thus impairing their operation,
- expose the heating element, which could then become overheated,
- result in steam being produced with possibly disastrous consequences.

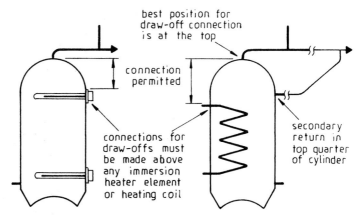

(a) Draw-off connections to hot water cylinders

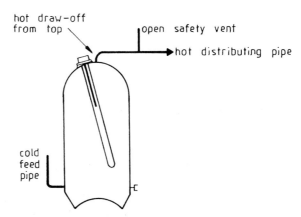

(b) Cylinder with top entry heater element

Where immersion heater or primary coil is inserted at top of cylinder any draw-off connection must be above immersion heater connection. The only exception is the use of a draining valve which must have a removable key for operation.

Figure 4.19   Control of water level

Chapter 5
Pipe sizing

Correct pipe sizes will ensure adequate flow rates at appliances and avoid problems caused by oversizing and undersizing, see figure 5.1.
Oversizing will mean:

- additional and unnecessary installation costs;
- delays in obtaining hot water at outlets;
- increased heat losses from hot water distributing pipes.

Undersizing may lead to:

- inadequate delivery from outlets and possibly no delivery at some outlets during simultaneous use;
- some variation in temperature and pressure at outlets, especially showers and other mixers;
- some increase in noise levels.

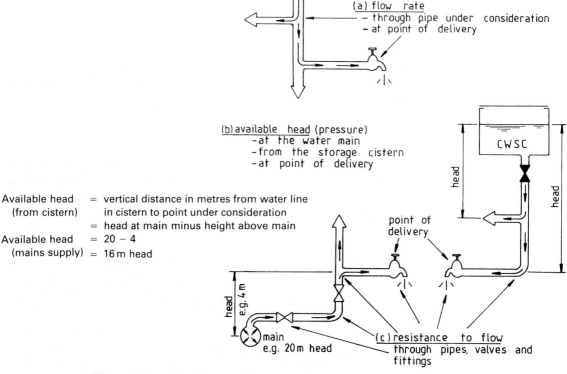

Available head = vertical distance in metres from water line
(from cistern) in cistern to point under consideration
= head at main minus height above main
Available head = 20 − 4
(mains supply) = 16 m head

Figure 5.1   Pipe sizing considerations

5.1 Sizing procedure for supply pipes

The procedure below is followed by an explanation of each step with appropriate examples.

(1) Assume a pipe diameter.
(2) Determine the flow rate:
  (a) by using loading units;
  (b) for continuous flows;
  (c) obtain the design flow rate by adding (a) and (b).
(3) Determine the effective pipe length:
  (d) work out the measured pipe length;
  (e) work out the equivalent pipe length for fittings;
  (f) work out the equivalent pipe length for draw-offs;
  (g) obtain the effective pipe length by adding (d), (e) and (f).
(4) Calculate the permissible loss of head:
  (h) determine the available head;
  (i) determine the head loss per metre run through pipes;
  (j) determine the head loss through fittings;
  (k) calculate the permissible head loss.
(5) Determine the pipe diameter:
  (l) Decide whether the assumed pipe size will give the design flow rate in (c) without exceeding the permissible head loss in (k).

Explanation of the procedure

Assume a pipe diameter

In pipe sizing it is usual to make an assumption of the expected pipe size and then prove whether or not the assumed size will carry the required flow.

Determine the flow rate

In most buildings it is unlikely that all the appliances installed will be used simultaneously. As the number of outlets increases the likelihood of them all being used at the same time decreases. Therefore it is economic sense to design the system for likely peak flows based on probability theory using loading units, rather than using the possible maximum flow rate.

(a) *Loading units*. A loading unit is a factor or number given to an appliance which relates the flow rate at its terminal fitting to the length of time in use and the frequency of use for a particular type and use of building (probable usage). Loading units for various appliances are given in table 5.1.

By multiplying the number of each type of appliance by its loading unit and adding the results, a figure for the total loading units can be obtained. This is converted to a design flow rate using figure 5.2.

An example using loading units is given in figure 5.3.

Table 5.1   **Design flow rates and loading units**

Outlet fitting	Flow rate l/s	Loading units
WC flushing cistern single or dual flush	0.10	2
WC trough cistern	0.15 per WC	2
Wash basin	0.15 per tap	$1\frac{1}{2}$ to 3
Spray tap or spray mixer	0.04 per tap	—
Bidet	0.15 per tap	1
Bath tap, nominal size $\frac{3}{4}$	0.30	10
Bath tap, nominal size 1	0.60	22
Shower head (will vary with type of head)	0.10 hot or cold	3
Sink tap, nominal size $\frac{1}{2}$	0.20	3
Sink tap, nominal size $\frac{3}{4}$	0.30	5
Washing machine or dishwasher	0.20 hot or cold	3
Urinal flushing cistern	0.004 per position served	—

Notes
(1) Flushing troughs are advisable where likely use of WCs is more than once per minute. For peak flows use 3 l/s.
(2) Mixer fittings use less water than separate taps, but this can be disregarded in sizing.
(3) Urinal demand is very low and can usually be ignored. Alternatively, use the continuous flow.
(4) Loading units should not be used for outlet fittings having high peak demands, e.g. those in industrial installations. In these cases use the continuous flow.

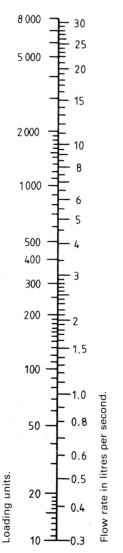

Figure 5.2   **Conversion chart – loading units to flow rate**

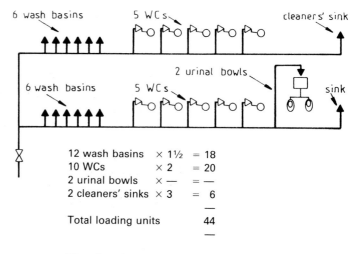

12 wash basins	× 1½	= 18
10 WCs	× 2	= 20
2 urinal bowls	× —	= —
2 cleaners' sinks	× 3	= 6
Total loading units		44

Therefore, from figure 5.2, the required flow rate for the system is 0.7 l/s.

Figure 5.3   **Example of use of loading units**

(b) *Continuous flows.* For some appliances, such as automatic flushing cisterns, the flow rate must be considered as a continuous flow instead of applying probability theory and using loading units. For such appliances the full design flow rate for the outlet fitting must be used, as given in table 5.1.

However, in the example shown in figure 5.3, the continuous flow for the two urinals of 0.008 l/s (from table 5.1) is negligible and can be ignored for design purposes.

(c) *Design flow rate.* The design flow rate for a pipe is the sum of the flow rate determined from loading units (a) and the continuous flows (b).

Determine the effective pipe length

(d) *Find the measured pipe length.* Figure 5.4 is an example showing how the measured pipe length is found.

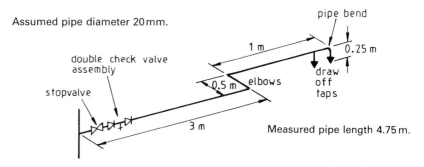

Assumed pipe diameter 20 mm.

Note There is no need to consider both branch pipes to taps.

Figure 5.4 **Example of measured pipe length**

(e) and (f) *Find the equivalent pipe lengths for fittings and draw-offs.* For convenience the frictional resistances to flow through fittings are expressed in terms of pipe lengths having the same resistance to flow as the fitting. Hence the term 'equivalent pipe length' (see table 5.2).

For example, a 20 mm elbow offers the same resistance to flow as a 20 mm pipe 0.8 m long.

Figure 5.5 shows the equivalent pipe lengths for the fittings in the example in figure 5.4.

Table 5.2   **Equivalent pipe lengths (copper, stainless steel and plastics)**

Bore of pipe mm	Equivalent pipe length			
	Elbow m	Tee m	Stopvalve m	Check valve m
12	0.5	0.6	4.0	2.5
20	0.8	1.0	7.0	4.3
25	1.0	1.5	10.0	5.6
32	1.4	2.0	13.0	6.0
40	1.7	2.5	16.0	7.9
50	2.3	3.5	22.0	11.5
65	3.0	4.5	—	—
73	3.4	5.8	34.0	—

Notes
(1) For tees consider change of direction only. For gate valves losses are insignificant.
(2) For fittings not shown, consult manufacturers if significant head losses are expected.
(3) For galvanized steel pipes in a small installation, pipe sizing calculations may be based on the data in this table for equivalent nominal sizes of smooth bore pipes. For larger installations, data relating specifically to galvanized steel should be used. BS 6700 refers to suitable data in the *Plumbing Design Guide* published by the Institute of Plumbing

Using the example from figure 5.4:

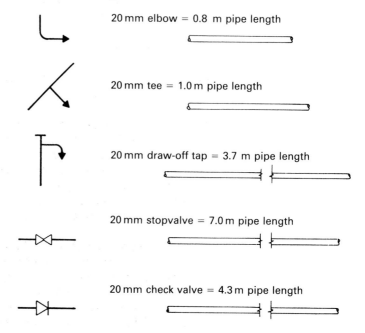

20 mm elbow = 0.8 m pipe length

20 mm tee = 1.0 m pipe length

20 mm draw-off tap = 3.7 m pipe length

20 mm stopvalve = 7.0 m pipe length

20 mm check valve = 4.3 m pipe length

Figure 5.5   **Examples of equivalent pipe lengths**

(g) *Effective pipe length.* The effective pipe length is the sum of the measured pipe length (d) and the equivalent pipe lengths for fittings (e) and draw-offs (f).

Therefore, in the example in figure 5.4 the effective pipe length would be:

Measured pipe length		4.75 m
Equivalent pipe lengths		
elbows	2 × 0.8 =	1.6 m
tee	1 × 1.0 =	1.0 m
stopvalve	1 × 7.0 =	7.0 m
taps	2 × 3.7 =	7.4 m
check valves	2 × 4.3 =	8.6 m
Effective pipe length		= 30.35 m

Permissible loss of head (pressure)

Pressure can be expressed in the following ways.

(i)   As force per unit area, N/m^2.
    $1 N/m^2 = 1$ pascal (Pa), which is the SI unit.

(ii) As a multiple of atmospheric pressure.
    Atmospheric pressure $= 100 kN/m^2 = 100 kPa = 1$ bar.

(iii) As metres head, that is, the height of the water column from the water level to the draw-off point.
    $1 m$ head $= 9.81 kN/m^2 = 9.81 kPa = 98.1 mb$

In the sizing of pipes, pressure is normally expressed in metres head.

(h) *Available head.* This is the static head or pressure at the pipe or fitting under consideration, measured in metres head (see figure 5.1).

(i) *Heat loss through pipes.* The loss of head (pressure) through pipes due to frictional resistance to water flow is directly related to the length of the pipe run and the diameter of the pipe. Pipes of different materials will have different head losses, depending on the roughness of the bore of the pipe and on the water temperature. Copper, stainless steel and plastics pipes have smooth bores and only pipes of these materials are considered in this section.

(j) *Head loss through fittings.* In some cases it is preferable to subtract the likely resistances in fittings (particularly draw-offs) from the available head, rather than using equivalent pipe lengths.
    Table 5.3 gives typical head losses in taps for average flows compared with equivalent pipe lengths. Figures 5.6 and 5.7 provide a method for determining head losses through stopvalves and float-operated valves respectively.

Table 5.3   **Typical head losses and equivalent pipe lengths for taps**

Nominal size of tap	Flow rate l/s	Head loss m	Equivalent pipe length m
$\frac{1}{2}$	0.15	0.5	3.7
$\frac{1}{2}$	0.20	0.8	3.7
$\frac{3}{4}$	0.30	0.8	11.8
1	0.60	1.5	22.0

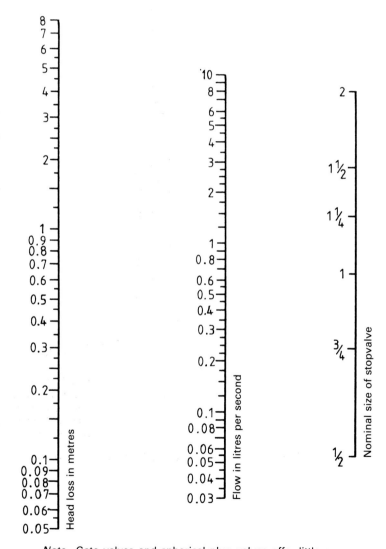

Note   Gate valves and spherical plug valves offer little or no resistance to flow provided they are fully open.

Figure 5.6   **Head loss through stopvalves**

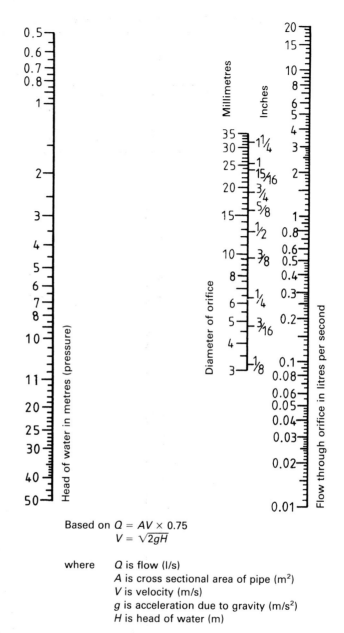

Based on $Q = AV \times 0.75$
$V = \sqrt{2gH}$

where Q is flow (l/s)
        A is cross sectional area of pipe (m^2)
        V is velocity (m/s)
        g is acceleration due to gravity (m/s^2)
        H is head of water (m)

Figure 5.7   Head loss through float-operated valves

(k) *Permissible head loss.* This relates the available head to the frictional resistances in the pipeline. The relationship is given by the formula:

$$\text{Permissible head loss (m/m run)} = \frac{\text{Available head (m)}}{\text{Effective pipe length (m)}}$$

This formula is used to determine whether the frictional resistance in a pipe will permit the required flow rate without too much loss of head or pressure. Figure 5.8 illustrates the permissible head loss for the example in figure 5.4.

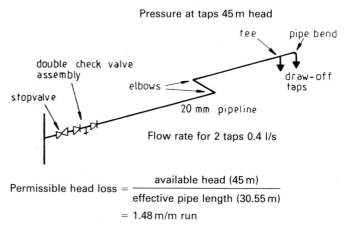

Pressure at taps 45 m head

double check valve assembly

stopvalve

elbows

20 mm pipeline

Flow rate for 2 taps 0.4 l/s

tee   pipe bend

draw-off taps

$$\text{Permissible head loss} = \frac{\text{available head (45 m)}}{\text{effective pipe length (30.55 m)}}$$

$$= 1.48 \, \text{m/m run}$$

Figure 5.8   **Example of permissible head loss**

Determine the pipe diameter

In the example in figure 5.4 a pipe size of 20 mm has been assumed. This pipe size must give the design flow rate without the permissible head loss being exceeded. If it does not, a fresh pipe size must be assumed and the procedure worked through again.

Figure 5.9 relates pipe size to flow rate, flow velocity and head loss. Knowing the assumed pipe size and the calculated design flow rate, the flow velocity and the head loss can be found from the figure as follows.

(i)   Draw a line joining the assumed pipe size (20 mm) and the design flow rate (0.4 l/s).
(ii) Continue this line across the velocity and head loss scales.
(iii) Check that the loss of head (0.12 m/m run) does not exceed the calculated permissible head loss of 1.48 m/m run.
(iv) Check that the flow velocity (1.4 ml/s) is not too high by referring to table 5.4.

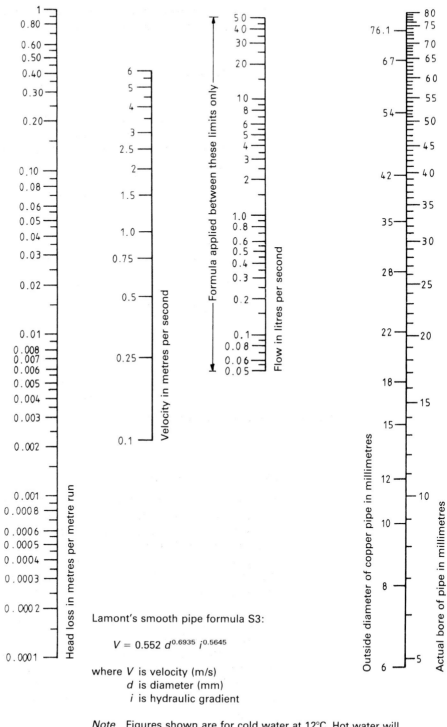

Lamont's smooth pipe formula S3:

$$V = 0.552\ d^{0.6935}\ i^{0.5645}$$

where V is velocity (m/s)
d is diameter (mm)
i is hydraulic gradient

Note Figures shown are for cold water at 12°C. Hot water will show slightly more favourable head loss results

Figure 5.9   Determination of pipe diameter

Table 5.4 **Maximum recommended flow velocities**

Water temperature °C	Flow velocity	
	Pipes readily accessible m/s	Pipes not readily accessible m/s
10	3.0	2.0
50	3.0	1.5
70	2.5	1.3
90	2.0	1.0

Note Flow velocities should be limited to reduce system noise.

5.2 Tabular method of pipe sizing

Pipe sizing in larger and more complicated buildings is perhaps best done by using a simplified tabular procedure. BS 6700 gives examples of this but for more detailed data refers to the Institute of Plumbing *Design Guide.*

The data used here in the tabular method are taken from BS 6700. I have varied the method to be more in line with the Institute of Plumbing *Design Guide.*

The tabular method uses a work sheet which sets out the steps to be followed in the pipe sizing procedure. An example of the method follows with some explanation of each step.

Explanation of the tabular method

Pipework diagram

(1) Make a diagram of the pipeline or system to be considered (see figure 5.10).
(2) Number the pipes beginning at the point of least head, numbering the main pipe run first, then the branch pipes.
(3) Make a table to show the loading units and flow rates for each stage of the main run. Calculate and enter loading units and flow rates, see figure 5.10.

Calculate flow demand

(1) Calculate maximum demand (see figure 5.10):
- add up loading units for each stage (each floor level);
- convert loading units to flow rates;
- add up flow rates for each stage.

(2) Calculate probable demand (see figure 5.10):
- add up loading units for all stages;
- convert total loading units to flow rate.

(3) Calculate percentage demand (number of stages for which frictional resistances need be allowed). See figure 5.12.

Work through the calculation sheet

See figure 5.11, using the data shown in figures 5.10 and 5.12.

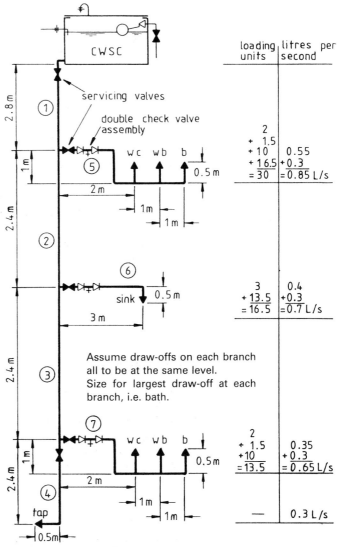

Bib tap at 0.3 l/s in frequent use.

Note Figure is not to scale for convenience, water level in cistern taken to be at base of cistern. Servicing valves assumed to be full flow gate valves having no head losses.

Refer also to figure 5.12

Figure 5.10   Pipe sizing diagram

This is an example of a suitable calculation sheet with explanatory notes.

Calculation sheet

(1) Pipe reference	(2) Loading units	(3) Flow rate (l/s)	(4) Pipe size (mm diameter)	(5) Loss of head (m/m run)	(6) Flow velocity (m/s)	(7) Measured pipe run (m)	(8) Equivalent pipe length (m)	(9) Effective pipe length (m)	(10) Head consumed (m)	(11) Progressive head (m)	(12) Available head (m)	(13) Final pipe size (mm)	(14) Remarks
Enter pipe reference on calculation sheet	Determine loading units (table 5.1)	Convert loading units to flow rate (figure 5.2)	Make assumption as to pipe size (inside diameter)	Work out frictional resistance per metre (figure 5.9)	Determine velocity of flow (figure 5.9)	Measure length of pipe under consideration	Consider frictional resistances in fittings (table 5.2 and figures 5.6 and 5.7)	Add totals in columns 7 and 8	Head consumed – multiply column 5 by column 9	Add head consumed in column 10 to progressive head in previous row of column 11	Record available head at point of delivery	Compare progressive head with available head to confirm pipe diameter or not	Notes

Note   If, for any pipe or series of pipes, it is found that the assumed pipe size gives a progressive head that is in excess of the available head, or is noticeably low, it will be necessary to repeat the sizing operation using a revised assumed pipe diameter.

Figure 5.11   Calculation sheet – explanation of use

Estimated maximum demand = 1.4 ℓ/s

Probable demand = 0.85 ℓ/s

$$\text{Percentage demand} = \frac{\text{Probable demand}}{\text{estimated maximum demand}} \times \frac{100}{1}$$

$$= \frac{0.85}{1.4} \times \frac{100}{1} = 60\%$$

Therefore only 60% of the installation need be considered.

For example, if we were designing for a multi-storey building 20 storeys high, only the first 12 storeys need to be calculated.

However, in the example followed here, the whole system has been sized because the last fitting on the run has a high flow rate in continuous use.

For branches only the pipes to the largest draw-off, i.e. the bath tap, need be sized.

Calculation sheet

(1) Pipe reference	(2) Loading units	(3) Flow rate (l/s)	(4) Pipe size (mm diameter)	(5) Loss of head (m/m run)	(6) Flow velocity (m/s)	(7) Measured pipe run (m)	(8) Equivalent pipe length (m)	(9) Effective pipe length (m)	(10) Head consumed (m)	(11) Progressive head (m)	(12) Available head (m)	(13) Final pipe size (mm)	(14) Remarks
1	30	0.85	32	0.05	1.2	2.8	1.4	4.2	0.21	0.21	2.8	32	
5	13.5	0.35	20	0.095	1.25	5.5	12.0	17.5	1.66	1.87	3.3	20	
2	16.5	0.7	25	0.12	1.5	2.4	—	2.4	0.29	2.16	5.2	25	
6	3	0.3	20	0.07	1.0	3.5	10.4	13.9	0.97	3.13	5.7	20	
3	13.5	0.65	25	0.1	1.4	2.4	—	2.4	0.24	3.37	7.6	25	
7	13.5	0.35	20	0.095	1.25	5.5	12.0	17.5	1.66	5.03	8.1	20	
4	—	0.3	20	0.07	1.0	2.9	1.6	4.5	0.31	5.34	10.0	20	

Refer also to figure 5.10

Figure 5.12   Calculation sheet – example of use

5.3 Sizing cold water storage

In Britain, cold water has traditionally been stored in both domestic and non-domestic buildings, to provide a reserve of water in case of mains failure. However, with new byelaws being introduced, we may see an increase in the use of 'direct' pressure systems.

BS 6700 is inconsistent about the quantity of storage required in houses. In clause 5.3.3 it recommends:

Smaller houses	cistern supplying cold water only	– 100 l to 150 l
	cistern supplying hot and cold outlets	– 200 l to 300 l
Larger houses	per bedroom	– 100 l

However, in clause 6.11.5 it recommends a minimum storage capacity of 230 l where the cistern supplies both cold water outlets and hot water apparatus, and this storage capacity was a requirement of byelaws in the past. I favour the old byelaw requirements, which are more specific and which installers are familiar with, i.e.:

Cold water storage cistern	– 115 l minimum
Feed cistern	– No minimum but should be equal to hot store vessel supplied
Combined feed and storage cistern	– 230 l minimum

For larger buildings the capacity of the cold water storage cistern depends on:

- type and use of building;
- number of occupants;
- type and number of fittings;
- frequency and pattern of use;
- likelihood and frequency of breakdown of supply.

These factors have been taken into account in table 5.5, which sets out minimum storage capacities in various types of building to provide a 24-hour reserve capacity in case of mains failure.

Table 5.5   **Recommended minimum storage of hot and cold water for domestic purposes**

Type of building	Minimum cold water storage l	Minimum hot water storage l
Hostel	90 per bed space	32 per bed space
Hotel	200 per bed space	45 per bed space
Office premises:		
with canteen facilities	45 per employee	4.5 per employee
without canteen facilities	40 per employee	4.0 per employee
Restaurant	7 per meal	3.5 per meal
Day school:		
nursery ⎱ primary ⎰	15 per pupil	4.5 per pupil
secondary ⎱ technical ⎰	20 per pupil	5.0 per pupil
Boarding school	90 per pupil	23 per pupil
Children's home or residential nursery	135 per bed space	25 per bed space
Nurses' home	120 per bed space	45 per bed space
Nursing or convalescent home	135 per bed space	45 per bed space

Note   The figures for hot water storage are based on those in the *Plumbing Design Guide* published by the Institute of Plumbing.

Example of calculation of minimum water storage capacity

Determine the amount of cold water storage required to cover 24 hours interruption of supply in a combined hotel and restaurant. Number of hotel guests 75, number of restaurant guests 350.

Storage capacity = number of guests × storage per person from table 5.5

Hotel storage capacity $= 75 \times 200 = 1500\,l$

Restaurant storage capacity $= 350 \times 7\ = 2450\,l$

Therefore total storage capacity required is $1500\,l + 2450\,l = 3950\,l$

5.4   Sizing hot water storage

Minimum hot water storage capacities for dwellings, from BS 6700, are:

- 45 l per occupant, unless the heat source provides a quick recovery rate,
- 100 l for systems heated by solid fuel boilers,
- 200 l for systems heated by off-peak electricity.

The feed cistern should have a capacity at least equal to that of the hot storage vessel.

The standard gives little information on storage capacities for larger buildings so in table 5.5 the author has given data for this based on that in the Institute of Plumbing Design Guide. The calculations are similar to those in section 5.3 for the minimum cold water storage capacity in larger buildings, and require the number of people to be multiplied by the storage per person.

To size the hot water storage for any installation, account must be taken of the following:

- pattern of use;
- rate of heat input to the stored water (see table 5.6);
- recovery period for the hot store vessel;
- any stratification of the stored water.

Table 5.6 **Typical heat input values**

Appliance	Heat input kW
Electric immersion heater	3
Gas-fired circulator	3
Small boiler and direct cylinder	6
Medium boiler and indirect cylinder	10
Directly gas-fired storage hot water heater (domestic type)	10
Large domestic boiler and indirect cylinder	15

Stratification (see figure 3.40) means that the hot water in the storage vessel floats on a layer of cold feed water. This enables hot water to be drawn from the storage vessel without the incoming cold feed water mixing appreciably with the remaining hot water. In turn, this allows a later draw-off of water at a temperature close to the design storage temperature, with less frequent reheating of the contents of the storage vessel and savings in heating costs and energy.

Stratification is most effective when cylinders and tanks are installed vertically rather than horizontally, with a ratio of height to width or diameter of at least 2:1.

The cold feed inlet should be arranged to minimise agitation and hence mixing, by being of ample size and, if necessary, fitted with a baffle to spread the incoming water.

Stratification is used to good effect in off-peak electric water heaters (see figure 5.13). In this case no heat is normally added to the water during the daytime use and consequently very little mixing of hot and cold water takes place. In other arrangements the heating of the water will induce some mixing.

(a) Bottom entry heater

more even temperature throughout cylinder

With a bottom entry immersion heater mixing will occur when the water is being heated.

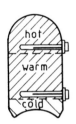

hot at top

warm

cold at bottom

(b) Top entry heater

With top entry immersion heater stratification will prevent mixing.

(c) Twin entry immersion heater

hot

warm

cold

With a twin entry immersion heater the top entry element can provide economical energy consumption for normal use with the bottom entry element operating when large quantities of water are needed, for example, for bathing and washing.

With off-peak heaters the top element can be brought into use during on-peak periods when needed to top up hot water, and stratification will ensure that on-peak electricity is not used to excess.

Figure 5.13   **Effects of stratification**

Calculation of hot water storage capacity

As noted above, the storage capacity required in any situation depends on the rate of heat input to the stored hot water and on the pattern of use. For calculating the required storage capacity BS 6700 provides a formula for the time M (in min) taken to heat a quantity of water through a specified temperature rise:

$$M = VT/(14.3P)$$

Where

V is the volume of water heated (in l);
T is the temperature rise (in °C);
P is the rate of heat input to the water (in kW).

This formula can be applied to any pattern of use and whether stratification of the stored water takes place or not. It ignores heat losses from the hot water storage vessel, since over the relatively short times involved in reheating water after a draw-off has taken place, their effect is usually small.

The application of this formula to the sizing of hot water cylinders is best illustrated by the following examples, in which figures have been rounded.

In these examples a small dwelling with one bath installed has been assumed. Maximum requirement: 1 bath (60 l at 60°C plus 40 l cold water) plus 10 l hot water at 60°C for kitchen use, followed by a second bath fill after 25 min.

Thus a draw-off of 70 l at 60°C is required, followed after 25 min by 100 l at 40°C, which may be achieved by mixing hot at 60°C with cold at 10°C.

Example 1   Assuming good stratification

Good stratification could be obtained, for example, by heating with a top entry immersion heater. With a rate of heat input of 3 kW, the time to heat the 60 l for the second bath from 10°C to 60°C is:

$$M = VT/(14.3P)$$
$$M = (60 \times 50)/(14.3 \times 3)$$
$$M = 70 \, min$$

Since the second bath is required after 25 min, it has to be provided from storage. But in the 25 min the volume of water heated to 60°C is:

$$V = M \, (14.3)/T$$
$$V = (25 \times 14.3 \times 3)/50$$
$$V = 21 \, l$$

Therefore the minimum required storage capacity is:

$$70 + 60 - 21 = 109 \, l$$

Example 2   Assuming good mixing of the stored water

Good mixing of the stored water would occur, for example, with heating by a primary coil in an indirect cylinder.

Immediately after drawing off 70 l at 60°C for the first bath and kitchen use, the heat energy in the remaining water plus the heat energy in the 70 l replacement at 10°C equals the heat energy of the water in the full cylinder.

The heat energy of a quantity of water is the product of its volume and temperature. Then, if V is the minimum size of the storage cylinder and T is the water temperature in the cylinder after refilling with 70 l at 10°C:

$$(V - 70) \times 60 + (70 \times 10) = VT$$
$$T = (60V - 4200 + 700)/V$$
$$T = (60V - 3500)/V$$
$$T = 60 - 3500/V$$

The second bath is required after 25 min. Hence, with a rate of heat input of 3 kW:

$$25 = VT/(14.3 \times 3)$$

and the temperature rise $\quad T = (25 \times 14.3 \times 3)/V$

and $\qquad\qquad\qquad\qquad T = 1072.5/V$

A temperature of at least 40°C is required to run the second bath. Therefore the water temperature of the refilled cylinder after the first draw-off of 70 l, plus the temperature rise after 25 min, must be at least 40°C, or:

$$(60 - 3500/V) + (1072.5V) = 40 \text{ (or more)}$$
$$60 - 2427.5/V = 40$$
$$20 = 2427.5/V$$
$$V = 122 \text{ l}$$

Table 5.7 **Hot water storage vessels — minimum capacities**

Heat input to water kW	Dwelling with 1 bath		Dwelling with 2 baths*	
	With stratification l	With mixing l	With stratification l	With mixing l
3	109	122	165	260
6	88	88	140	200
10	70	70	130	130
15	70	70	120	130

Note * Maximum requirement of 130 l drawn off at 60°C (2 baths plus 10 l for kitchen use) followed by a further bath (100 l at 40°C) after 30 min.

These calculations, which may be carried out for any situation, show the value of promoting stratification wherever possible. They also show the savings in storage capacity that can be made, without affecting the quality of service to the user, by increasing the rate of heat input to the water.

5.5 *Legionella* – implications in sizing storage

Recent investigations into the cause and prevention of *Legionella* suggest that hot and cold water storage should be sized to cope with peak demand only, and that in the past, storage vessels have often been over-sized. For example, this book follows BS 6700 in sizing cold water storage to provide for 24 hours interruption of supply.

Reduced storage capacities would mean quicker turnover of water and less opportunity for *Legionella* and other organisms to flourish.

Stratification in hot water can lead to ideal conditions for bacteria to multiply, as the base of some cylinders often remain within the 20°C to 50°C temperature range. The base of the cylinder may also contain sediment and hard water scale which provides an ideal breeding ground for bacteria.

Where, however, good mixing occurs, higher temperatures can be obtained at the cylinder base during heating up periods. This is helpful because *Legionella* will not thrive for more than five minutes at 60°C and are killed instantly at temperatures of 70°C or more.

Chapter 6
Preservation of water quality

Water undertakers in England and Wales have a duty under the Water Act 1989 to provide a supply of wholesome water which is suitable and safe for drinking and culinary purposes. Similar conditions apply in Scotland. At the same time, public demand requires that water supplied is of good appearance with minimal colour, taste or odour.

To enable water undertakers to maintain their supplies in wholesome condition and to preserve the quality of water supplied, the Act provides water undertakers with powers to enforce their water byelaws which came into effect on 1 January 1989. Installers and users must ensure that systems and components which are installed comply with water byelaws.

BS 6700 looks at the preservation of water quality in four main areas.

(1) *Materials in contact with water*. There is not much point in having a good quality water supply if it becomes contaminated by unsuitable pipes, fittings and jointing materials.
(2) *Stagnation of water* and the prevention of bacterial growth particularly at temperatures between 20°C and 50°C.
(3) *Cross connections* must be prevented between pipes supplied directly under mains pressure and pipes supplied from other sources, such as:
  (i)   water from a private source
  (ii) non-potable water
  (iii) stored water
  (iv) water drawn off for use.
(4) *Prevention of backflow* from fittings or appliances into services or mains, particularly at point of use, i.e. draw-off taps, flushing cisterns, washing machines, dishwashers, storage cisterns and hose connections. For example, pumps should not be connected so as to cause backflow into the supply pipe, (see figure 6.1).

Backflow prevention forms the largest portion of clause 9 of BS 6700 and is prominent in water byelaws, which aim to implement the recommendations of the *Backsiphonage Report* of 1974.

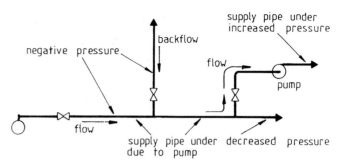

Pump may cause pressure drop sufficient to reverse flow in adjacent pipes.

Figure 6.1   **Example of backflow caused by pump installation**

6.1   Materials in contact with water (Part II of the water byelaws)

Contamination of water by contact with unsuitable materials will be avoided if careful attention during design and installation is given to:

- the specification and selection of acceptable materials used in the manufacture of pipes, fittings and appliances;
- the method of installation, and in particular to the method of and materials used when jointing and connecting pipes, fittings and appliances;
- the environment into which pipes, fittings and appliances are to be installed;
- the design of the various elements of installation, especially where differing materials are to be used.

Materials selected for use in contact with water intended for domestic purposes should comply with water byelaws and in particular Part II *Prevention of contamination of water from contact with unsuitable materials or substances*. In general this means compliance is assured where any materials used are manufactured to a relevant British Standard specification, or are listed in the *Water Fittings and Materials Directory* produced under the United Kingdom Water Fittings Byelaws Scheme.

Examples of problems caused by unsuitable materials, and situations which should be avoided are shown in figures 6.2 and 6.4.

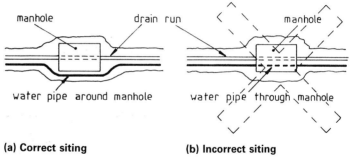

(a) Correct siting   **(b) Incorrect siting**

Pipes not to pass through any foul soil, refuse or refuse chute,
ash pit, sewer or drain, cess pool or manhole (Byelaw 5).

Figure 6.2   **Pipes not to pass through manholes, etc.**

Pipes made of any material which is susceptible to permeation by gas
or to deterioration by contact with substances likely to cause contamination of water should not be laid or installed in a place where permeation
or deterioration is likely to occur. Because plastic pipes in particular are
liable to a degree of permeation and deterioration by gas and oil, care
should be taken when positioning pipelines to avoid contact in the event
of leakages occurring from oil or gas lines, for example, at petrol filling
stations. Figure 6.3 shows the positioning of pipelines in trenches as
suggested by the National Joint Utilities Group Report No. 6.

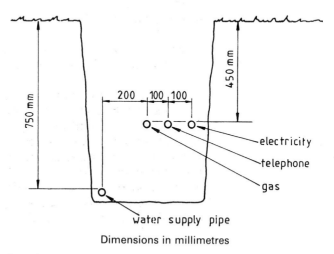

Dimensions in millimetres

Figure 6.3   **Positioning of pipelines in trenches**

Byelaw 7 refers to pipe materials or other materials which must not cause contamination of water used for domestic purposes.

Substances leached from some materials may adversely affect the quality of water. Although British Standard schemes for the testing of pipes, fittings and materials aim to prevent this, much depends on the installer and the material chosen for a particular application, after taking due account of the nature of the water (see figure 6.4).

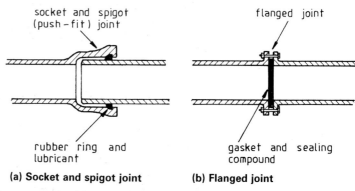

socket and spigot (push-fit) joint   flanged joint

rubber ring and lubricant   gasket and sealing compound

(a) Socket and spigot joint   **(b) Flanged joint**

See table 11.14 for jointing materials and guidance on their use.

The *Water Fittings and Materials Directory* produced by the Water Research Centre lists materials and fittings which are approved for use.

Figure 6.4   **Examples of materials which could cause contamination if not chosen with care**

The use of lead for pipes, cisterns or in solders is prohibited because of the obvious danger of plumbo solvency (Byelaw 9). However, BS 6700 does give details of jointing for lead pipes within its list of jointing methods for potable water pipework because there will be cases when connections have to be made to existing lead pipes (see section 11.14).

For soldered joints to copper pipes, lead-free solders should be specified, e.g. tin/silver alloys.

The use of coal tar for the lining of pipes and cisterns is also prohibited. The direct connection between copper and lead pipes which might lead to electrolytic action and further lead solvency is prohibited in the absence of suitable means to prevent corrosion though galvanic action.

6.2 Stagnation of water

To restrict bacterial growth in stored drinking water, temperatures should be maintained as follows:

- cold water should be stored and distributed at as low a temperature as is practical and preferably below 20°C

- hot water should preferably be stored at 60°C to 65°C and distributed at not less than 50°C. (*Legionella* will not flourish at temperatures above 50°C)

Research into Legionnaires' Disease is ongoing. Evidence of this disease occurring has generally been found where pipes and components are not regularly maintained and cleaned, or in parts of systems that cannot easily be cleaned during routine maintenance programmes, e.g. where flushing valves or drain valves are wrongly positioned, or where accumulation of rust and other debris can settle and build-up. Infection by *Legionella* can be prevented through good design, and regular, thorough maintenance. Advice should be sought from the Department of Health for up-to-date information.

6.3   Prevention of contamination by cross connection

No supply pipe, distributing pipe or cistern used for conveying or receiving water supplied by a public water supplier shall be connected so that it can receive or convey water which is not supplied by a public water supplier, except where the water supplied by a public water supplier is discharged through a type A air gap (Byelaw 13 (2)), (see figure 6.5).

(a) Connections between supply pipe and distributing pipe

No connection shall be made between a supply pipe and a distributing pipe (Byelaw 13).

Figure 6.5   Cross connection hazards

continued

(b) Supplies from different sources

Installation not permitted. Correct installation is type A air gap on supply pipe.

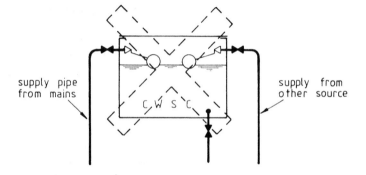

supply pipe from mains

supply from other source

C W S C

(c) Temporary connection to sealed system

No closed circuit to be connected to a supply pipe.

Important because primary circuits to heating systems are often heavily contaminated with additives.

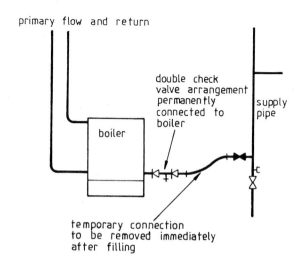

primary flow and return

double check valve arrangement permanently connected to boiler

supply pipe

boiler

temporary connection to be removed immediately after filling

(d) Shower mixer installation

No connection to be made between a supply pipe and a hot or cold distributing pipe (Byelaw 13).

Water could flow from hot store vessel to the drinking tap if the supply pipe should fail.

May be permitted if feed cistern is protected to Byelaw 30 standards and a check valve is fitted to both the hot and cold supplies to the shower.

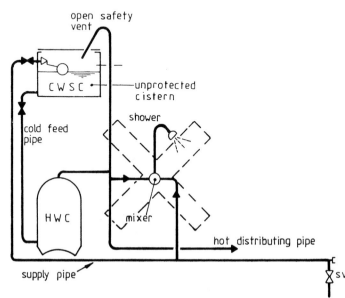

open safety vent

C W S C

unprotected cistern

shower

cold feed pipe

H W C

mixer

hot distributing pipe

supply pipe

sv

Figure 6.5   Cross connection hazards
continued

Additives to primary hot water or heating circuits

If a liquid (other than water) is used in any type of heating primary circuit, or if an additive, e.g. a corrosion inhibitor, is used in water in such a circuit, the liquid or additive should be non-toxic and non-corrosive.

6.4 Backflow protection

Backflow can include both 'backsiphonage' and 'back pressure'. Appropriate preventative measures should be related to the level of risk, should backflow occur, and will vary according to the nature and use of the supply.

The *Backsiphonage Report* of 1974 divides the levels of risk into three categories.

Class 1   Risk of serious contamination occurring continuously or frequently, and likely to be harmful to health.
Class 2   Risk of contamination by substances which may be harmful to health, but not continuously or frequently present.
Class 3   Risk of contamination by substances not likely to be harmful to health.

The Department of the Environment *Byelaws Guidance Notes* list appliances that fall into each category, and also give relevant prevention devices. BS 6700 has similar lists. Examples of backflow risk are shown in table 6.1 whilst a more comprehensive list is shown in tables 6.4, 6.5 and 6.6.

Table 6.1   Examples of backflow risk

Source of risk	Examples of recommended protection
Class 1 risk WC pan Bidet	Approved flushing cistern correctly installed Type A air gap
Class 2 risk Sink taps Hose union taps Washing machine	Type A air gap recommended Double check valve assembly Type B air gap
Class 3 risk Mixer tap Water softener (domestic)	Any of the above devices, or a check valve

Backflow prevention devices

Type A air gap

See figures 6.6 and 6.7. This is a pipe arrangement where the supply to a cistern or vessel is arranged as follows:

- the cistern or vessel has an unrestricted overflow to atmosphere;
- the supply discharge and its outlet is not obstructed;
- the water is discharged vertically or not more than 15° from the vertical;
- the vertical distance between the discharge pipe outlet and the spill-over level of the cistern or vessel is *not less* than that shown in table 6.2.

Its principal merit is that is has no moving parts and cannot readily be destroyed by vandals. It is accepted throughout the world.

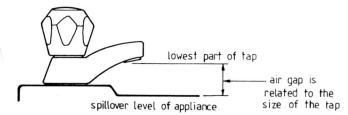

Type A air gap to be visible, measurable and unobstructed.

Figure 6.6   **Example of use of type A air gap at draw-off tap**

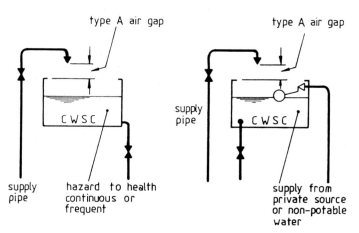

(a) Cistern with hazardous contents   **(b) Cistern with mixed supplies**

Air gap related to size of inlet (see table 6.3).

Flow from inlet to be into air at atmospheric pressure and not more than 15° from the vertical.

Figure 6.7   **Examples of type A air gap at cisterns**

Table 6.2   **Air gaps at taps**

Nominal size of tap or fitting	Vertical distance between tap outlet and spill-over level of receiving appliance mm
Up to and including $\frac{1}{2}$	20
Over $\frac{1}{2}$ and up to and including $\frac{3}{4}$	25
Over $\frac{3}{4}$	70

Table 6.3   **Dimensions of type A air gaps to cisterns** (dimensions in mm)

Bore diameter of feed pipe or outlet	Vertical distance of point of outlet above spill-over level
Up to 14	20
Over 14 up to 21	25
Over 21 and up to 41	70
Over 41	Twice the bore of the feed pipe or outlet

Type B air gap

See figure 6.8. In a cistern or other similar vessel, which is open at all times to the atmosphere, the vertical distance between the lowest point of discharge and the critical water level should be one of the following:

(1) sufficient to prevent backsiphonage of water into the supply pipe;
(2) not less than the distances shown in table 6.3.

In storage cisterns the type B air gap, as shown in figure 6.8 (a), is impractical. Its critical water level, which determines the difference in fixing heights between the float operating valve and the warning pipe, cannot be readily calculated. The critical water level will vary from one installation to another depending on the inlet pressure and the length and gradient of the warning pipe.

In most cases it is expected that water undertakers will accept the simpler arrangement seen in figure 6.8 (b).

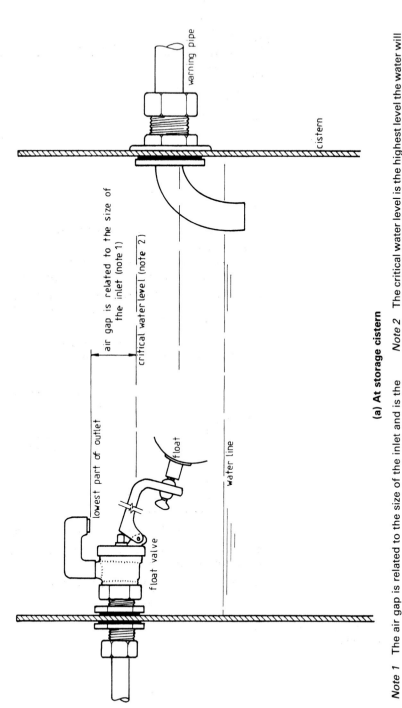

(a) At storage cistern

Note 1 The air gap is related to the size of the inlet and is the minimum permitted vertical distance between the 'critical' water level and the lowest part of the float valve outlet (see table 6.3).

Note 2 The critical water level is the highest level the water will reach at the maximum rate of inflow, i.e. float removed.

Figure 6.8   Type B air gap

continued

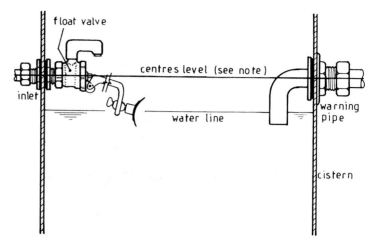

(b) Acceptable alternative to the type B air gap

This arrangement is acceptable if:
- ○ the cistern complies with Byelaw 30;
- ○ the float operated valve is of the reducing flow type.

A reducing flow type float valve is one which gradually closes as the water level in the cistern rises, e.g. diaphragm float valve to BS 1212: Parts 2 or 3.

In this cistern the critical water level is assumed to be level with the centre line of the float valve body.

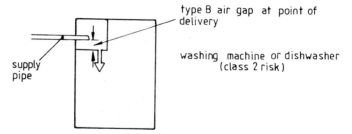

(c) Example of type B air gap to domestic appliance

Water Research Centre approved appliance will have protection device built in during manufacture.

Type B air gap to comply with BS 6281: Part 2.

Discharge to be unobstructed.

Figure 6.8 **Type B air gap**
continued

Pipe interrupter

A pipe interrupter (see figure 6.9) will admit air into a system, without the use of moving or flexible parts, to prevent backflow of water when a vacuum occurs. Whilst this device contains no moving parts, it can be subject to vandalism (by blockage of the airways).

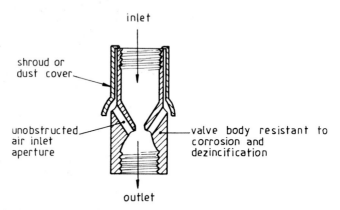

Alternative to type B air gap.

Valve should produce vacuum on outlet side.

Must be fitted at least 300 mm above overflowing level of appliance served.

No tap or valve to be installed downstream (outlet side).

Pipe downstream not to be reduced in size.

Length of pipe downstream to be as short as possible.

Pipe interrupter to be readily accessible for repair.

Pipe interrupter to comply with BS 6281: Part 3.

Figure 6.9   Example of pipe interrupter

Anti-vacuum valve or vacuum breaker

An anti-vacuum valve or vacuum breaker (see figure 6.10) is a mechanical device with an air inlet which remains closed when water flows past it, but which opens to admit air if there is a vacuum in the pipe.

The vacuum breaker must close once pressure and flow return to normal.

However, despite being mentioned in water byelaws as a vacuum breaking device, there are doubts as to whether it conforms to the byelaw requirements. It is suggested that they do not close effectively after use.

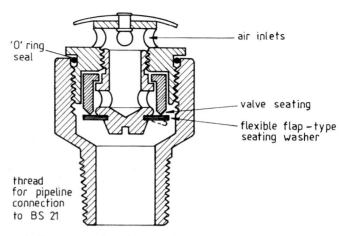

Anti-vacuum valves in some instances may discharge water from the air inlet valve while working normally. Therefore they must only be used where they can readily be seen and where no damage can occur from any leakage.

Diagram shows terminal type; in-line type also available.

Needs regular opening and shutting to keep it operable.

Must be same size as pipe to which it is connected.

Failure of these valves to close would mean waste of water and possible damage to building.

Satisfies Byelaw 11 if it complies with BS 6282: Part 3 and there is no control valve downstream.

Must be fitted at least 300 mm above overflow level of appliance served.

Figure 6.10 **Vacuum breaker, atmospheric type to BS 6282: Part 3**

Upstands are required for anti-vacuum valves to work to maximum effect, and provide additional protection should the protection at point of use fail, (see figure 6.11).

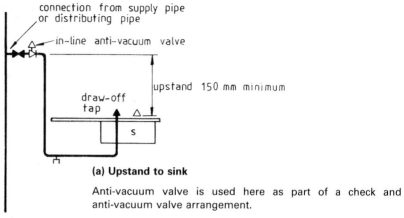

connection from supply pipe
or distributing pipe

in-line anti-vacuum valve

upstand 150 mm minimum

draw-off
tap

s

(a) Upstand to sink

Anti-vacuum valve is used here as part of a check and anti-vacuum valve arrangement.

Upstand is necessary for anti-vacuum valve to be effective.

Upstand 150 mm for cisterns and fixed appliances; 300 mm in other cases.

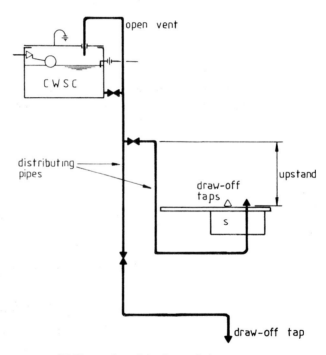

open vent

CWSC

distributing pipes

upstand

draw-off taps

s

draw-off tap

(b) Upstand to sink, cistern fed

Open vent shown here is an alternative to the check and anti-vacuum valve.

Upstand is needed for the vent to be fully effective.

Where vent is used, upstand should be at least 300 mm.

Branch not to rise above its connection.

Figure 6.11   Upstands to anti-vacuum valves

continued

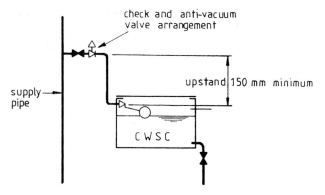

(c) Upstand to cistern

Upstand 150 mm above cistern and fixed appliances; 300 mm in other cases.

Upstand 150 mm above cistern and fixed appliances; 300 mm in

Figure 6.11 continued   **Upstands to anti-vacuum valves**

Check valve

A check valve (see figure 6.12) is a mechanical device which, by means of a resilient 'elastic' seal or seals, permits flow of water in one direction only and is closed when there is no flow.

The check valve should:
 - be resistant to corrosion and dezincification;
 - operate satisfactorily at temperatures up to 65°C;
 - when shut, prevent any flow from inlet to outlet where water pressure does not exceed 10 mb.

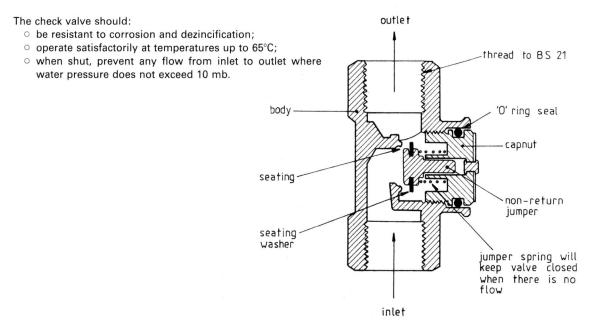

Figure 6.12   Check valve to BS 6282: Part 1

Double check valve assembly

This comprises either:

(1) two check valves with a test cock fitted between them (see figure 6.13(a));

or

(2) a similar arrangement to that in figure 6.13(a) produced as a single fitting (see figures 6.13(b) and 6.14).

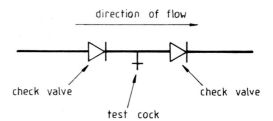

(a) Two check valves with a test cock between them

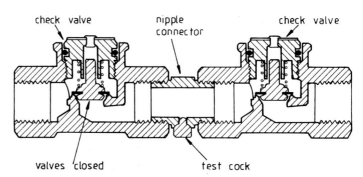

(b) Double check valve assembly produced as a single fitting

Figure 6.13   **Double check valve assembly**

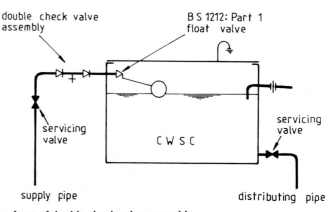

Figure 6.14   **Example of use of double check valve assembly**

Combined check and anti-vacuum valve

These are shown in figures 6.15 and 6.16.

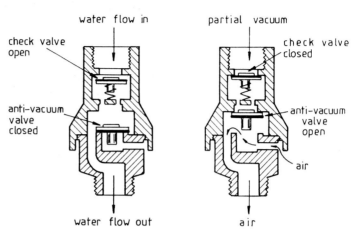

Must be used with an upstand for anti-vacuum valve to be effective.

Figure 6.15   **Combined check and anti-vacuum valve assembly**

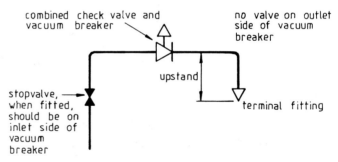

Upstand 150 mm above cistern and fixed appliances; 300 mm in other cases.

Check valve to be fitted upstream of vacuum breaker.

Figure 6.16   **Use of combined check valve and anti-vacuum valve assembly**

Backflow risks in bidets

Bidets are an obvious contamination risk that come within the Class 1 category of backflow risk. There are two types of bidet: the over rim type, and the ascending spray or submerged inlet type. Each is protected differently. The former may be used with its taps under mains pressure, the latter is required to be supplied from a storage cistern.

Over rim types should be arranged with air gaps in accordance with table 6.2 between the tap outlets and the spill-over level of the appliance (see figure 6.17). It is an offence to attach a hand-held flexible spray or similar fittings to taps or bidets where supply is from the mains.

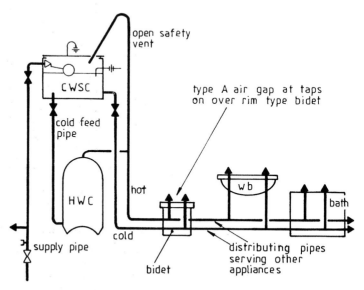

Over rim type bidets may be supplied from distributing pipe or supply pipe provided type A air gap is maintained.

Figure 6.17   **Over rim type bidet**

Ascending spray types are not permitted to be connected directly to a supply pipe. There are two acceptable ways of making connections to this type of bidet:

* by separate distribution pipes on both hot and cold, which must not be connected to any other appliance, except for WCs or urinals on the cold distributing pipe (see figure 6.18);
* by various break cistern arrangements (see figure 6.19).

Similar conditions apply to bidets with hand-held sprays (whether or not they are over rim types).

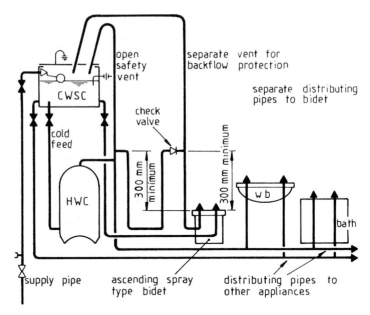

(a) Separate distributing pipes bidet

Precautions required to prevent backflow from bidet through cylinder to appliances:

- vent to atmosphere;
- check valve fitted downstream of vent;
- 300 mm minimum upstand.

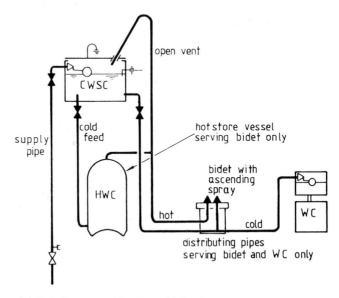

(b) Hot store vessel feeding bidet only

Connection between WC and bidet permitted because this creates no additional hazard.

No special precautions required where hot store vessel feeds the bidet only.

Figure 6.18   Ascending spray type bidet

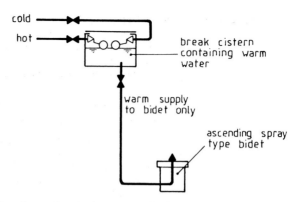

Supplies to break cistern may be from storage or direct from mains, but not from mixed sources.

Water in break cistern is likely to cool down between uses.

Figure 6.19   Example of break cistern arrangement to bidet

Categories of backflow risk

Tables 6.4, 6.5 and 6.6 list appliances which fall into each of the three backflow risk categories.

Table 6.4   Class 1 backflow risks

Examples of points of use or delivery of water where backflow is, or is likely to be, harmful to health from a substance continuously or frequently present. Note that the required protection is a type A air gap or an interposed cistern.
WC pan Urinal bowl Bedpan washer Dental sputum bowl Water treatment plant other than water softening plant Fire sprinkler systems containing anti-freeze Any appliance or cistern that can receive water not supplied by a public water supplier, e.g. supplementary sources of supply Sealed central heating system other than in dwellings Hose union tap not on domestic premises* Poultry and animal drinking troughs Installations in laboratories, dairies, slaughter houses, butchery and meat trade premises, dye works and sewage works Bottle washing apparatus Cisterns connected to central heating systems other than in dwellings Agricultural storage cisterns

* Hose union tap may be fitted with a double check valve or combined check valve and vacuum breaker as an alternative, but only with written permission of the water undertaker

Table 6.5 **Class 2 backflow risks**

Examples of points of use or delivery of water where backflow or
backsiphonage is, or is likely to be, harmful to health from a substance
which may be present, but not continuously or frequently.

Note that the protection acceptable under water byelaws is a type A or type B
air gap, a pipe interrupter, a combination of a check valve and a vacuum
breaker, or a double check valve assembly.

If an appropriate upstand cannot be provided, a combination of check valve
and vacuum breaker is unacceptable because vacuum breakers do not
afford reliable protection against back pressure backflow. Also, if the
vacuum breaker is of the atmospheric type, no control valve is permissible
downstream.

Vented primary hot water circuits and associated cisterns in dwellings
Home dialysing machine with integral membrane washing facility
Hose union tap on domestic premises (e.g. in kitchen, garage or garden)
Shower hose where shower head could be submerged in associated bath,
basin, etc.
Common-salt regenerated water softening plant in any premises other than
a single dwelling
Permanent standpipe supplying water to small boats in marinas, etc.
Temporary standpipe on mains, supplying water to mobile apparatus,
construction sites, etc.
Temporary connection to supply pipe of sealed primary circuits in single
dwellings
Clothes washing machine, dishwasher or tumble drier connected
permanently or temporarily to a water service (only type B air gap or pipe
interrupter acceptable)*
Drink vending or dispensing machine in which any ingredient or gas is
injected under pressure
WC cistern not supplied through a BS 1212: Part 2 or 3 float valve installed
with its centre line no lower than the critical water level

---

* Washing machines, dishwashers and tumble driers should be supplied from
storage when installed in premises other than a dwelling.

Table 6.6 **Class 3 backflow risks**

Examples of points of use or delivery of water where backflow or
backsiphonage is not, or is not likely to be, harmful to health.

Note that a check valve would be the acceptable protection.

Common-salt regenerated water softener in single dwelling
Home dialysing machine without integral membrane washing facility
Fire sprinkler system without storage connected to a supply pipe
Hot and cold water single outlet mixing valve or tap in a single dwelling with
supplies at differing pressures; prevention devices to be fitted to both hot
and cold feeds in addition to appropriate backflow protection at mixed
water discharge
Drink vending or dispensing machine in which no ingredient or gas is
injected under pressure

6.5 Secondary backflow protection

This is used where there is an increased risk of backflow because of the following:

- exceptionally heavy use;
- presence of a hazardous substance;
- possibility of internal backflow in buildings of multiple occupation, e.g. flats or tall buildings.

Secondary backflow protection should be provided in addition to prevention devices installed at points of use (as in tables 6.4, 6.5, and 6.6) and is required when:

- supply or distributing pipes convey water to two or more separately occupied premises;
- premises are required to provide a storage cistern capacity for 24 or more hours of normal use.

Acceptable arrangements for secondary backflow protection are shown in figure 6.20.

(a) Secondary backflow protection on supply pipes

Double check valve preferred.

No need for secondary protection at lowest level.

No part of the branch pipe to be higher than its connection to the common distributing pipe.

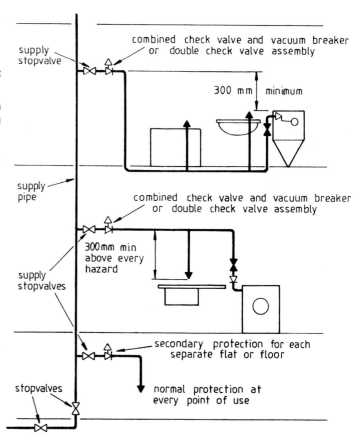

Figure 6.20   Acceptable arrangements for secondary backflow protection

continued

(b) Secondary backflow protection on distributing pipes

No part of the branch pipe to be higher than the distributing pipe.

Double check valve assembly preferred.

No need for secondary protection at lowest level.

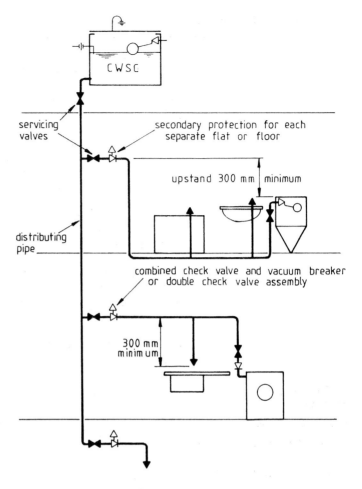

(c) Secondary backflow protection on distributing pipes using a vent pipe

Vent will admit air to prevent backflow

No need for secondary protection at lowest level

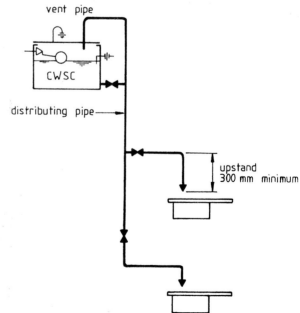

Figure 6.20   Acceptable arrangements for secondary backflow protection continued

6.6 Additional backflow protection

There are many industrial and trade premises where risk of contamination may be present, and where backflow protection may be required in addition to any secondary backflow protection that is provided. Examples of some of these are shown in the following pages, and include improvements to backflow protection in existing fire sprinkler systems.

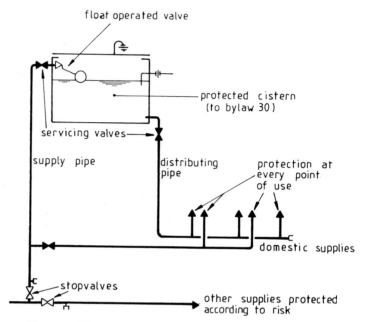

Pipes used for domestic purposes should not be connected to pipes for any other purpose.

Figure 6.21   Domestic supplies to dwellings and other premises

Figure 6.22 shows the prevention of contamination by backflow or cross connection within industrial, commercial, trade, research, educational, medical and similar establishments, in addition to any secondary backflow protection.

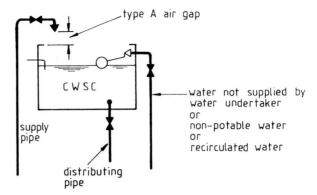

(a) Water supplied by undertaker and non-potable water from other sources

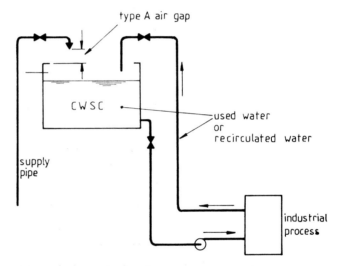

(b) Re-used or recirculated water

This connection is permitted only if distributing pipe remains under gravity flow from cistern. Otherwise, type A air gap must be used.

If distributing pipe supplies a hazardous situation, such as a pathology laboratory, no other fitting must be supplied from it.

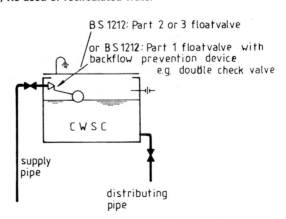

(c) Distribution pipe under gravity flow at all times

Figure 6.22   Non-domestic supplies

continued

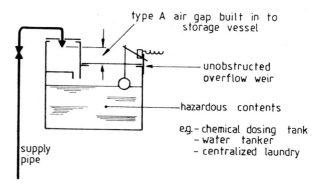

(d) Supplies to fixed or mobile appliances in industrial processes

Figure 6.22   **Non-domestic supplies**
continued

In premises where drinking water and non-potable water supplies are made available, they should be clearly identified as 'drinking water', 'non-potable water', or 'fire-fighting water', as appropriate. Hoses intended for drinking water should be used only for that purpose and should be marked 'not for cleaning purposes'.

Figure 6.23 shows acceptable methods for the prevention of contamination of mains by backflow or cross connection in ports and similar establishments, (in addition to any secondary backflow protection).

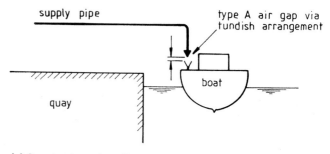

(a) Supply pipe taken directly to ship or boat

Figure 6.23   **Prevention of backflow in ports and similar establishments**

continued

(b) Supply taken by gravity from elevated storage

No pump must be fitted to this supply. Otherwise type A air gap must be used at storage cistern.

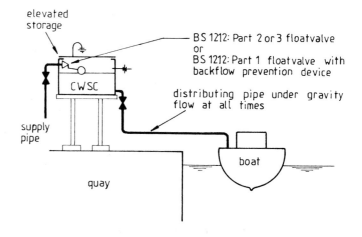

(c) Pumped supply from quayside storage cistern

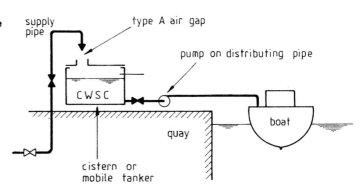

(d) Supplies from standpipe through hose of small diameter

Where hose is larger than 22 mm, anti-backflow device must be at least 1 m above deck level

Pipes and mechanical devices to be protected from frost.

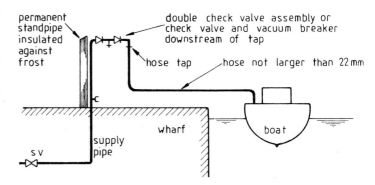

Figure 6.23   Prevention to backflow in ports and similar establishments
continued

Figure 6.24 shows the prevention of contamination by backflow or cross connection within agricultural and similar establishments (in addition to any secondary backflow protection).

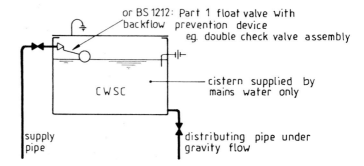

(a) Supplies from storage

Where pump is fitted to distributing pipe, cistern must be fitted with type A air gap.

Supplies to static or mobile appliances, e.g. crop sprayer, to be connected via type A air gap or properly connected break cistern serving no other appliance.

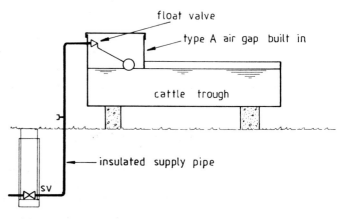

(b) Supplies to cattle troughs

Figure 6.24   **Prevention of backflow in agricultural establishments**   *continued*

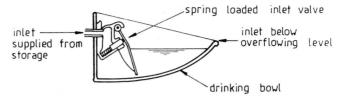

(c) Supply to animal drinking bowl from storage

If inlet has type A air gap and is shrouded from mouth contact, the drinking bowl could be supplied direct from mains.

Further backflow protection needed if other appliances are supplied from same source.

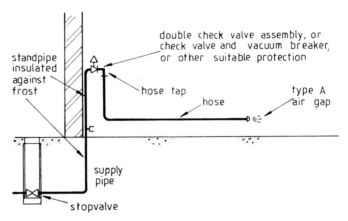

(d) Supplies by hose pipe direct from supply pipe

Type A air gap to be maintained at end of hose, or hose must be supplied via correctly connected break cistern.

Figure 6.24 continued **Prevention of backflow in agricultural establishments**

Protection for fire sprinkler systems

The scope of this book does not permit detailed discussion of fire sprinkler systems, which are subject to Rules of the Loss Prevention Council and the requirements of the local water undertakers who should be consulted before commencing any installation.

Sprinkler systems are commonly installed, particularly in high-risk situations. They are fitted for emergency fire protection only and should be used for no other purpose. Water in them may become stagnant and create contamination risks particularly where substantial volumes are stored in ground level or elevated cisterns.

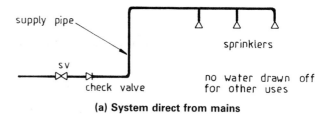

(a) System direct from mains

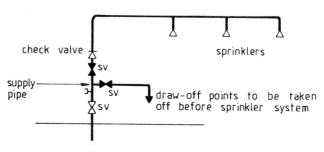

(b) Wet system with draw-offs

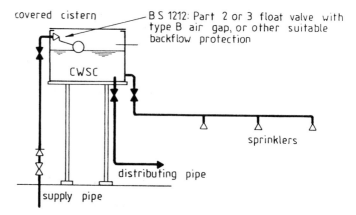

(c) Wet system from storage

Cistern supplied from mains only.

Water for other purposes to be drawn separately from storage cistern.

An uncovered cistern must supply the sprinklers only.

Figure 6.25   Backflow protection in fire sprinkler systems

continued

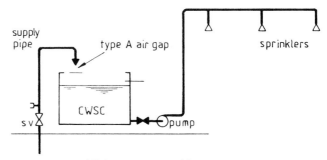

(d) System pumped from storage

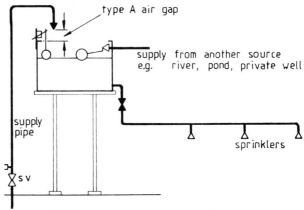

(e) System supplied from mixed sources

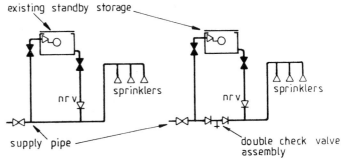

(f) Improving existing systems

An alternative to the double check valve arrangement would be a pressure principle backflow preventor. The best solution, provided the effectiveness of the sprinkler system is not reduced, would be to remove the connection between the supply pipe and the sprinkler distributing pipe.

This installation only applies to the improvement of protection to existing buildings. It will not be permitted in new installations.

Figure 6.25 **Backflow protection in fire sprinkler systems**
continued

Chapter 7
Frost precautions

When water loses heat and its temperature drops below freezing point it turns to ice. Upon freezing its volume increases by approximately 10%. This results in:

- damage to pipework and fittings due to the greater volume of ice compared to water;
- the risk of explosion should hot water apparatus be put into operation when part of the system is blocked with ice.

The temperature of water supplied is quite low. In winter, just a small reduction in temperature will cause freezing.

The best precaution against freezing of water services in buildings is the obvious one of keeping the inside of the building continuously warm by the provision and maintenance of adequate heating. This will be easier to achieve and more economical if the building has been designed and constructed so as to minimise the loss of heat through its structure. When the whole building is not heated, or where heating is only intermittent, the localised heating of water pipes and fittings or the heating of their immediate surroundings may suffice. Localised heating, such as trace heating, in conjunction with a frost thermostat, should be used only in addition to other forms of frost protection, or where those are unsuitable.

7.1 Location of pipes

Pipes should be located so as minimise the risk of frost damage, avoiding areas that are difficult to keep warm, especially the following:

- unheated parts of the building, e.g. roof spaces, cellars and underfloor spaces, garages or outhouses;
- areas where draughts could occur, e.g. near windows, air bricks, ventilators, external doors or under suspended floors;
- cold surfaces, e.g. chases and ducts in outside walls and places where pipes come into direct contact with outside walls;
- any exterior position above ground.

If it is not possible to avoid these locations then protection should be provided.

Although some plastics pipes and cisterns are flexible and not easily fractured by ice formation, such formation stops the supply of water. Therefore precautions are still necessary.

7.2   Protection of pipes and fittings

Pipes underground

These should be laid at least 750 mm below the ground surface bearing in mind any expected changes in ground levels (see figure 7.1). This minimum byelaw requirement will be sufficient in most cases, but where more severe weather can be expected, then greater depth of cover can be provided up to a maximum of 1350 mm.

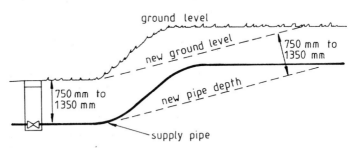

Where ground is to be levelled, ensure pipe is maintained at full depth of cover. This is important, especially on new installations, where ground is often levelled after services are laid.

Minimum depth of cover must be maintained over whole length of pipe.

Figure 7.1   Pipes in uneven ground

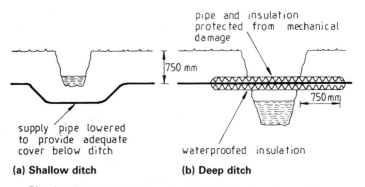

Pipe insulated where impractical to take it below deep ditch.

Figure 7.2   Pipes passing under or through ditches

The insulating value of the soil will be affected by its nature and water-retaining capacity as well as the degree of exposure of the site. Therefore extra depth of cover may be required to suit local conditions, see figure 7.2.

Stopvalves below ground

Underground stopvalves should not be brought up to a higher level merely for ease of access, but should remain at the same depth as the pipe.

Pipes entering buildings

Pipes should enter buildings at the same depth as laid (see figures 7.3 and 7.4). Where it is not possible to maintain a minimum depth of 750 mm, pipes should be insulated.

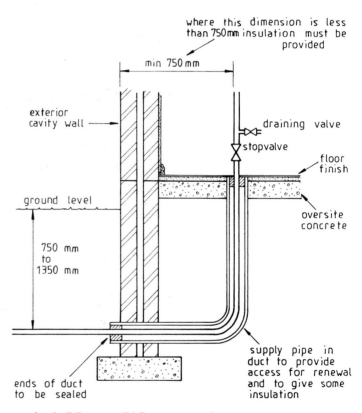

Figure 7.3   Pipes entering buildings – solid floor construction

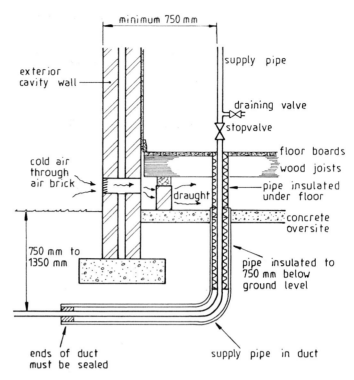

Shallow foundations as shown here would not be normal in new buildings.

Figure 7.4   **Pipes entering buildings – suspended timber floor**

Pipes and fittings outside buildings

Where possible these should be laid underground. Where this is not possible sufficient protection against the effects of frost should be provided. See figures 7.5 and 7.6.

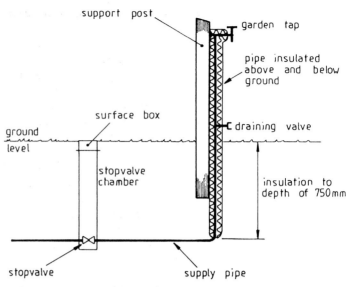

Insulation to be waterproofed throughout and protected from damage.

Note Backflow protection device needed if hose pipe is to be connected.

Figure 7.5   Pipe rising from below ground to garden stand-pipe

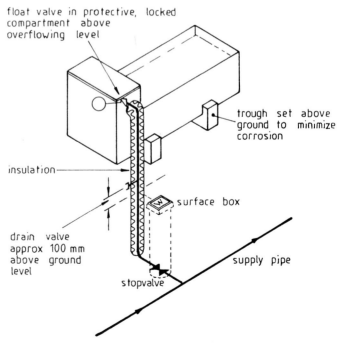

Insulation above and below ground, waterproofed and protected from damage by animals.

Figure 7.6   Pipe rising from below ground to cattle trough

Pipes and cisterns in or above roof spaces

When considering the insulation of roof spaces and pipes and components within roof spaces (see figure 7.7), full advantage should be taken of
any heat rising from rooms. Roof spaces are notoriously cold and
draughty places and the cause of many frozen pipes in the UK. This has
not been helped by new Building Regulations which require ventilation of
roof spaces to prevent condensation. Consideration should also be given
to the following:

- ceiling insulation should be omitted from below cisterns;
- insulation should be provided all over any cistern and include any
  pipes rising to connect with it;
- pipes within roof spaces, including warning and overflow pipes, should
  wherever possible be fixed below any ceiling insulation;
- any pipes other than those situated as above should be adequately
  insulated.

If a cistern is sited above the roof of a building, it must be protected by
installing it, together with its inlet and outlet pipes, in an insulated
enclosure provided with either its own means of heating, or opening into
some heated part of the building itself.

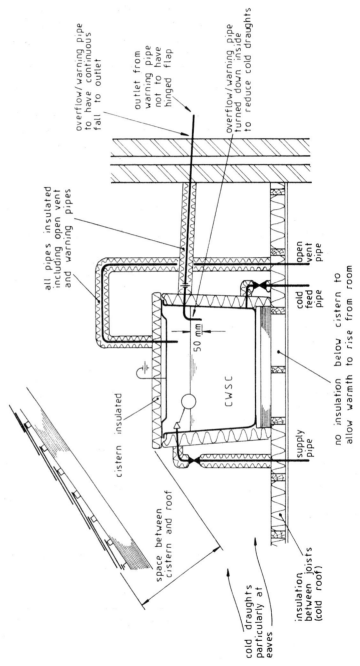

Figure 7.7 Insulation of pipes and cisterns in roof space

7.3 Insulation

The minimum thicknesses of thermal insulating materials used for the protection of water pipes and fittings are shown in table 7.1.

Space should be allowed around pipes for the insulation which should be continuous over pipes and fittings with space left only for the operation of valves.

Where necessary, insulating material must be resistant to, or should be protected from, mechanical damage, rain, moist atmosphere, subsoil water and vermin.

Manufacturers' literature will give details of application and use of insulating materials in various situations.

Note that no amount of insulation can prevent a pipe from freezing. It can, however, slow down the heat loss and delay the effects of frost.

Table 7.1   Minimum thickness of thermal insulating material to delay freezing for frost protection

Nominal outside diameter of pipe mm	Thermal conductivity of insulating material not exceeding:							
	0.035 W/m °C	0.04 W/m °C	0.055 W/m °C	0.07 W/m °C	0.035 W/m °C	0.04 W/m °C	0.055 W/m °C	0.07 W/m °C
	Indoor installations				Outdoor installations			
	mm	mm	mm	mm	mm	mm	mm	mm
Up to and including 15	22	32	50	89	27	38	63	100
Over 15 up to and including 22	22	32	50	75	27	38	63	100
Over 22 up to and including 42	22	32	50	75	27	38	63	89
Over 42 up to and including 54	16	25	44	63	19	32	50	75
Over 54 up to and including 76.1	13	25	32	50	16	25	44	63
Over 76.1 and flat surfaces	13	19	25	38	16	25	32	50

Notes
(1)   Pending revision of BS 5422 this table lists the thermal conductivity value with an air temperature of 0°C and the minimum thickness of insulating material providing protection against freezing during normal occupation of buildings.
(2)   Storage cisterns and pipework in roof spaces are usually considered as indoor installations.
(3)   Pipework in the air space beneath a suspended ground floor or in a detached garage should be protected as outdoor installations.

Table 7.2   Examples of insulating materials

Thermal conductivity W/m°C	Material
Less than 0.035	Polyurethane foam Foamed or expanded plastics including rigid and flexible preformed pipe insulation of these materials
0.04 to 0.055	Corkboard
0.055 to 0.07	Exfoliated vermiculite (loose fill)

7.4 Draining facilities

It should be possible to drain down all parts of a hot or cold water system including its pipes, fittings, components and appliances, so that pipes not in use can be emptied in cold weather (see figure 7.8). An empty pipe cannot suffer frost damage. For pipes to be adequately drained, allowance must be made for the entry of air. Usually this will be done through taps and vent pipes but in some special cases air inlet valves will be needed.

To assist draining:

- pipes should be laid to slight and continuous falls to draining valve at low points;
- draining taps should have means for connection of a hose pipe;
- cisterns, cylinders and tanks should be fitted with draining taps unless they can be drained through pipes leading to a draining tap elsewhere.

Draining taps must be of the screw-down pattern having a removable key and must comply with BS 2879 (see figure 7.9).

BS 6700 recommends that all pipes and fittings be drainable. However, it is my view that there are some exceptions to this rule. Draining taps should not be fitted within sealed ducts or below floors or below ground where they may be inaccessible. More importantly, they should not be fitted where they might become submerged and create a contamination risk. Figures 7.10 and 7.11 show examples of draining tap positions.

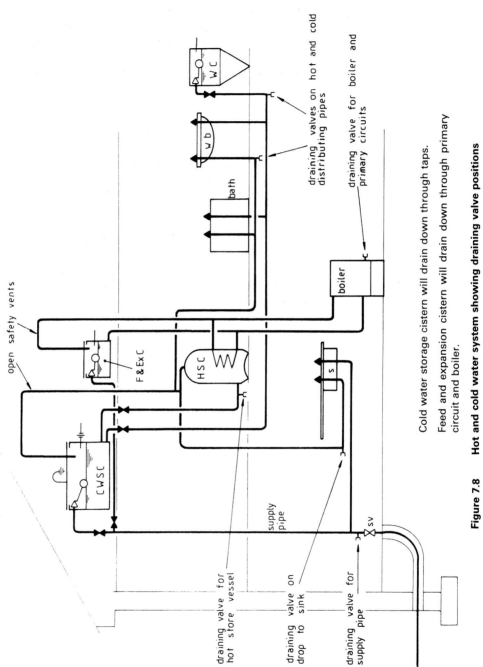

Cold water storage cistern will drain down through taps.

Feed and expansion cistern will drain down through primary circuit and boiler.

Figure 7.8   Hot and cold water system showing draining valve positions

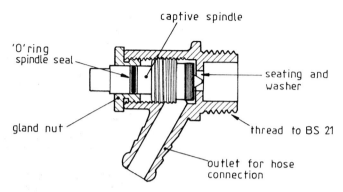

Figure 7.9 **Drain cock to BS 2879**

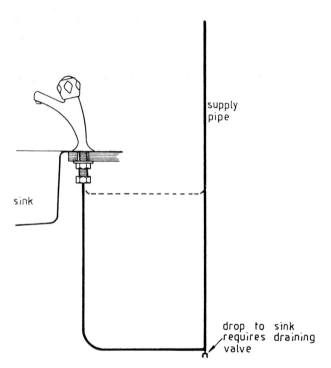

Some water undertakers may permit very short, shallow drops
(see dotted line) to be considered drainable through the tap.

Figure 7.10 **Draining arrangements – pipe drops to draw-offs**

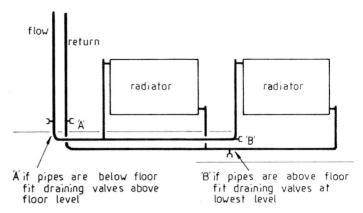

'A' if pipes are below floor
fit draining valves above
floor level

'B' if pipes are above floor
fit draining valves at
lowest level

For pipe drops below floor level where the floor is likely to be
covered by carpets and furniture it may be better to fit the
draining valves above, but near the floor (alternative A).

Alternative B is more suitable where pipes do not drop below
floor level.

Figure 7.11   **Draining arrangements – pipe drops below floors**

Chapter 8
Water economy and energy conservation

BS 6700 suggests a relationship between water usage and energy consumption, based on the cost of supplying water and disposing of effluent. It also suggests that metering will produce considerable savings in water consumption.

There is a need to conserve water, because as we consume more, sources become increasingly costly and difficult to find. It is therefore in the interest of the consumer, and of concern to the plumber, that water is not consumed unduly.

With regard to metering and the cost of water supply, I only partly agree with the standard's philosophy. Whilst metering will most likely help to reduce the consumption of water, the cost to the consumer must increase because of the difficulties of reading and maintaining meters, and the additional cost of purchasing and installing meters and ancillary materials. Under the present system consumers on higher rateable values who opt for meters, do receive smaller water bills but, it is feared, at the expense of the majority whose properties are on lower rateable values.

If metering becomes widespread or mandatory, then the costs of installing, maintaining and reading meters will be enormous and must in the long run cause a general increase in payments for water. At the same time, the energy consumed in manufacturing, installing and maintaining meters and their associated components is likely to exceed the energy savings from reduced water consumption.

8.1   Water economy

Water byelaws have always been written in order to prevent or reduce wastage or excessive use of water. The new byelaws which came into effect on 1 January 1989 are no exception. Water is costly to produce, and during hot dry spells it is becoming increasingly difficult to maintain a constant supply. BS 6700 has been written with byelaws and water economy in mind and suggests a number of ways in which savings can be made.

Leakages

Approximately 30% of all the potable water produced in Britain is lost through waste, undue consumption or misuse. Much of this is due to leakages underground from mains and service pipes. Water undertakers constantly carry out waste detection programmes to reduce this loss, but because of the age and condition of many of our underground pipes, it is a continuing problem which can only be solved by regular testing and monitoring of supply pipes and systems. (See chapter 12 which includes methods for the detection of leakages in metered and unmetered supply pipes.)

Additionally, BS 6700 suggests that:

(1) Warning pipes from cisterns should discharge where they can readily be seen (Byelaw 1), e.g. over a doorway where the discharge may cause a nuisance.
(2) Ponds and pools should be built so that water loss is kept to a minimum. They should not lose more than 3 mm depth of water per day, after taking rainfall and evaporation into consideration. Ponds and pools of more than 10,000 l capacity must be fitted with an impervious lining or membrane (Byelaw 47).

Flushing of WC cisterns and urinals

WC flushing accounts for about 25% of all domestic water used in buildings. Urinals consume water in large quantities if not properly controlled.

To reduce the amount of water used in flushing, byelaws now require that the sizes of cisterns used in future for the flushing of WCs and urinals shall in due course (1993) be reduce compared with those in previous byelaws. This has led to considerable debate, particularly when considering the self-cleansing of drains.

WC flushing cisterns and troughs are required to be of the valveless type incorporating siphonic action. Flushing valves are not permitted.

Cisterns, when flushed, must be capable of emptying the contents of the pan effectively using a single flush.

The capacity of cisterns supplying water to new WCs will be reduced to a maximum of 7.5 l per flush (see figures 8.1 and 8.2) in 1993. 7.5 litre apparatus is currently available and accepted by the water industry. Dual-flush type cisterns will be permitted only for replacement after 1993.

WC pans will need to be designed to suit the associated cistern to ensure effective cleansing. Take care that the correct cistern or pan is fitted when either is renewed.

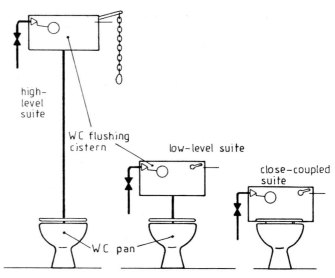

Maximum permitted cistern capacity:
- single flush, 7.5 l;
- dual flush, 9.5 l (not permitted after 1 January 1993, except for replacement).

Figure 8.1   **Capacity of WC flushing cisterns**

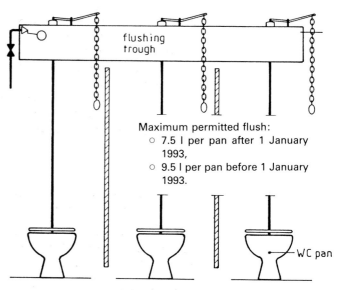

Maximum permitted flush:
- 7.5 l per pan after 1 January 1993,
- 9.5 l per pan before 1 January 1993.

Figure 8.2   **Capacity of WC trough cisterns**

Table 8.1   **Maximum permitted capacity of WC cisterns**

Type	Use	Volume (Scotland)		Dates and restrictions
Single flush	Domestic	7.5		Effective from 1 January 1989 (Scotland from 1 January 1993)
Dual flush	Domestic	9.5		Not permitted after 1 January 1993
Single flush	Non-domestic	7.5	(9.5)	9.5 l until 1 January 1993
Dual flush	Non-domestic	9.5		Not permitted after 1 January 1993
Flushing trough	Anywhere	7.5 per pan		9.5 l until 1 January 1993

Note   Existing cisterns may be replaced by similar cisterns after the above dates.

For urinal flushing, it is recommended that automatic flushing cisterns should supply a maximum of 2.5 l per flush per bowl, stall, or 700 mm length of slab, delivered not more than three times per hour (see figure 8.3).

Additionally, supply pipes to urinal flushing cisterns are required to be fitted with devices that will control the supply during periods when the urinal is not in use. There are two methods of controlling the supply: time control (see figure 8.6); and hydraulic control (see figures 8.7 and 8.8).

Where a time control is used, the pipe supplying the urinal cistern should also be fitted with a lockable control valve.

Alternatively, individual bowls or stalls may be user operated as required, e.g. by a chain pull (see figure 8.4). A multiple-style urinal installation is shown in figure 8.5.

BS 6700 differs a little from the new water byelaws but, except for individual appliances, means virtually the same (Byelaws 82 and 83). Byelaw requirements for the maximum flushing volumes of water to be used in the flushing of urinals are shown in table 8.2.

It may not always be feasible to size the automatic flushing cistern for multiple urinals according to the number of bowls, stalls or slabs, or to give the age-old recommended three flushes per hour.

Table 8.2   **Maximum permitted volumes of water for flushing urinals**

Appliance	Up to 1 January 1989	After 1 January 1989
For a single bowl or stall	No more than necessary	10 l/h
For more than one bowl, stall, or per 700 mm width of slab	15 l/h	7.5 l/h

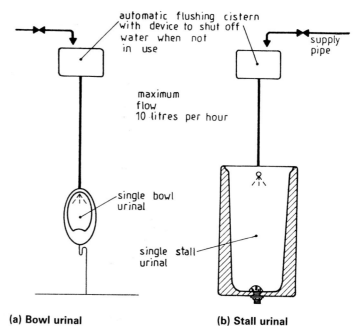

(a) Bowl urinal   (b) Stall urinal

Figure 8.3   Urinal flushing – single appliances

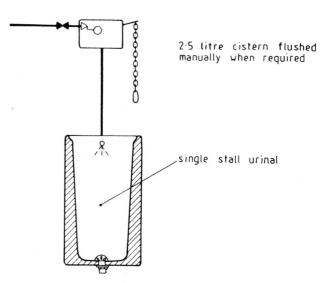

Figure 8.4   Single urinal with manual flushing control

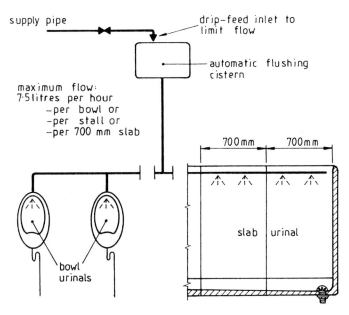

Supply pipe to be fitted with device to shut off water when not in use.

Figure 8.5   Multiple urinal installation

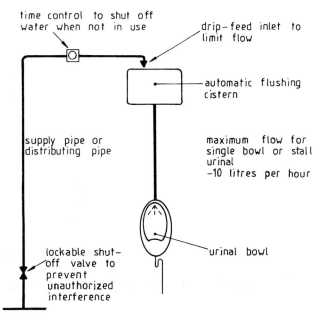

Figure 8.6   Time control to automatic flushing cistern

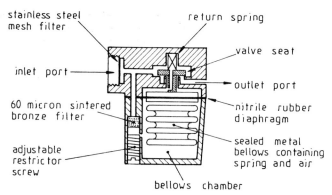

Figure 8.7   **Hydraulic flow control device for automatic flushing cistern**

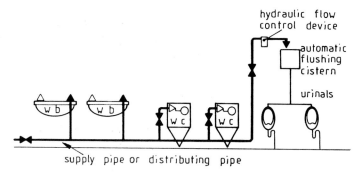

No use of appliance = no flow through flow control device.

Draw-offs at other appliances will create pressure variations causing the hydraulic flow control to open, thus permitting a small quantity of water to pass into the automatic flushing cistern.

Figure 8.8   **Use of hydraulic flow control device**

Table 8.3 provides for a variation of flushing arrangements based on the formula:

$$\text{Time interval (min)} = \frac{\text{cistern capacity (l)} \times \text{time in minutes (60)}}{\substack{\text{maximum flush requirement} \\ \text{(l) (Byelaws 82 and 83)}} \times \substack{\text{number of bowls, stalls} \\ \text{or 700 mm width of slab}}}$$

For example, using a 13.5 l cistern to flush five urinal bowls

$$\text{Time interval} = \frac{\text{cistern capacity (13.5 l)} \times \text{time (60)}}{\text{maximum flush (7.5 l per bowl)} \times \text{number of bowls (5)}}$$

$$= \frac{13.5 \times 60}{7.5 \times 5}$$

$$= 21.6 \, \text{min}$$

Table 8.3   **Volumes and flushing intervals for urinals**

Number of bowls, stalls, or per 700 mm of slab	Volume of automatic flushing cistern l				Maximum fill rate l/h
	4.5 l	9 l	13.5 l	18 l	
	Shortest period (minutes) between flushes				
1	27	54	81	108	10
2	18	37	54	72	15
3	12	24	36	48	22.5
4	9	18	27	36	30
5	7.2	14.4	21.6	28.8	37.5
6	6	12	18	24	45

The reduction in quantity of water used should lead to the reduction of limescale build up in appliances and drains, and consequent reduction in odours.

Waste plugs

Waste plugs should be fitted to all baths, basins, sinks or similar appliances, except where delivery is less than 3.5 l/min and the appliance is designed not to have a plug, e.g. basins with spray taps, and shower trays. Appliances for medical or veterinary purposes are also excepted.

Self-closing taps

Self-closing taps should be of the non-concussive type (see figure 8.9) and be capable of closing against 2.6 times the working pressure. These taps are very effective when new, but tend to fail in the open position after a period in use so they should only be used in buildings where regular maintenance and inspection can be ensured.

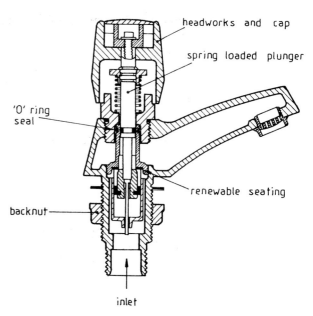

Figure 8.9   **Non-concussive self-closing tap**

Spray taps

Spray taps can provide savings of up to 50% in both fuel and water. However, they have several disadvantages:

- they require regular maintenance;
- they are only suitable for hand rinsing;
- the heads may block in hard water areas;
- since self-cleansing velocities of waste water may not be achieved, residues may build up in waste pipes.

Aerators

Aerators may reduce consumption and when compared with spray taps will give improved flow pattern.

Showers

Showers are generally said to use less water than baths. However, the reduced consumption is often offset by more frequent use. Also, with the arrival of pressurised hot water systems, water consumption for showering will increase.

Washing troughs and fountains

Fittings serving these appliances should be capable of discharging to individual units without discharging to other units. A unit means a 600 mm length of straight trough or of the perimeter of a round appliance.

Domestic appliances

Maximum amounts of water used per complete cycle of operations:

Dishwasher:	7 l per place setting
Clothes washer without water-using tumbler dryer:	3.6 l/l of drum or tub volume
Clothes washer with water-using tumbler dryer:	6.4 l/l of drum volume
Water-using tumbler dryer:	20 l/kg dry load

Other economy measures

The following measures may also improve water economy:

- protection from mechanical damage and corrosion especially underground (trench preparation and backfill, pipe depth);
- use of British Standard or Water Research Centre approved fittings, components and pipes;
- adequate frost precautions;
- use of spray taps.

8.2 Energy conservation

Hot water storage vessels must be fitted with adequate thermal insulation. Insulating materials should limit heat loss to 90 W/m^2 of surface area of the storage vessel. The hot store vessel should be fitted with a thermostat to keep the water at the required temperature, and a time switch that will shut off the heat source when there is no demand. Also energy can be saved by reducing the quantity of water heated, bearing in mind the methods of heating or controlling the heat input to the storage vessel.

Pipes to and from hot water apparatus and central heating components should be insulated unless they are designed to contribute to space heating. Insulation provided should have a thermal conductivity value not exceeding 0.045 W/m/°C. The thickness of insulation should equal the pipe diameter, up to a maximum of 40 mm. Alternatively pipe insulation may meet the recommendations of BS 5422. Table 3.1 gives maximum lengths permitted for hot water distributing pipes without insulation. However, it is recommended that all pipes for hot water supply be lagged.

The use of pumps for both hot and cold water systems should be minimised and in consultation with the water undertaker, consideration should be given to the prudent use of mains water pressure. The energy needed to boost water pressure in a building is about 0.02 kWh/m^3 per metre of lift. Therefore it is advisable, where mains pressure is insufficient to supply the upper floors of a building, to use mains pressure to the limit of its supply.

Chapter 9
Noise and vibration

Noise is caused by vibration. In water systems many materials are susceptible to vibration and will transmit or even accentuate noises produced.

Most system noises, which many operatives explain away as unavoidable, can be avoided by better design and workmanship.

9.1   Flow noises

Pipework noise becomes significant at water velocities over 3 m/s. It is important therefore that systems are designed to keep water velocities below 3 m/s by increasing the pipe size, as necessary. The causes of some common flow noises are shown in figures 9.1, 9.2 and 9.3.

friction, vibration and noise where water rubs against pipe walls

Where velocities are below 3 m/s noise is not significant.

Figure 9.1   Flow noise in pipes

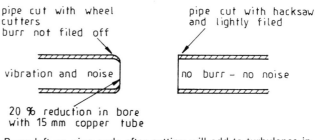

pipe cut with wheel cutters burr not filed off

vibration and noise

20 % reduction in bore with 15 mm copper tube

pipe cut with hacksaw and lightly filed

no burr — no noise

Burrs left on pipe ends after cutting will add to turbulence in pipes and increase noise levels.

Figure 9.2   Effect on water flow of burrs on pipe ends

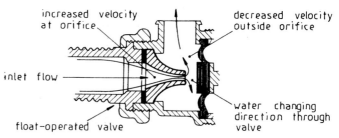

Changes in flow direction, and drop in pressure through valve may lead to cavitation which will further increase flow noise. Can be reduced by lowering flow velocity.

Figure 9.3   **Noise caused by water flow through narrow apertures**

Cavitation

Cavitation can be simply described as wear or erosion of the internal surfaces of pipes and fittings caused by turbulent water flow.

When cavitation occurs, water flow noise increases. Cavitation is not common in pipework as it usually only occurs at water velocities of 7 m/s to 8 m/s in elbow fittings. However, it can occur through reduced pressures at the upper parts of systems which incorporate long pipe drops.

Outlet fittings generally incorporate abrupt changes of direction in water flow, and there is a sudden drop in pressure at the outlet side of the seating of taps and float valves. These conditions are ideal for cavitation, and are the major cause of noise in these fittings. Although figure 9.4 shows a bib tap to BS 1010, this is a problem common to most types of taps and valves.

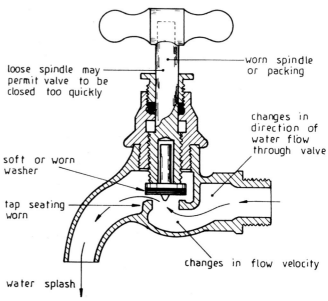

Figure 9.4   **Causes of noise in taps and valves**

Cavitation noises can be controlled by reducing the pressure drop across the valve seating. For example, if the inlet pressure is reduced and the tap is opened more fully it will operate more quietly. Similarly, a float valve with a lower inlet pressure and larger orifice will operate with less noise.

Fast flow through narrow apertures

Noises in float valves and stopvalves can often mean that the waterway is becoming blocked with particles carried along in the water (see figure 9.5). To avoid this, stopvalves should always be left in the fully open position so that particles can pass through and out of the system.

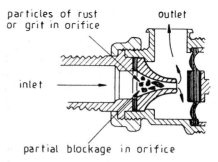

Partial blockage means smaller aperture and increase in velocity, vibration and noise.

Any obstruction in a valve which significantly reduces the bore will create noise. Commonly found in:
○ stopvalves that are partially closed;
○ stopvalves that are partially blocked;
○ float valves partially blocked (as illustrated).

It is the mistaken belief of many that to partially close a stopvalve will reduce pressure. In fact, it will create noise and unnecessary wear to the valve.

Figure 9.5   **Noise caused by obstruction in valves**

In the past, a good way to reduce the problem of blockage, and also reduce general flow noise in float valves, was to fit an equilibrium ballvalve. The use of the portsmouth type valve is restricted under the new water byelaws, but there is a diaphragm type shown in the approved water fittings list. They may also be permitted if used in conjunction with a double check valve assembly fitted down stream.

9.2   Water hammer noise

Valve closure

Sudden valve closure will cause shock waves to be transmitted along pipes with a loud hammer-ring noise. Rapid closure can be prevented by regular maintenance, making sure that packing glands are correctly adjusted and spindles are not loose. Other precautions include the restriction of velocities and avoidance of long straight pipe runs. Limiting water velocities to 3 m/s will not, in itself, reduce water hammer, but will help reduce the magnitude of pressure peaks produced.

Solenoid valves and self-closing taps

These often cause water hammer noise. Use non-concussive types, properly and regularly maintained (see figure 8.9).

Vibration of the cistern wall

This is very common in copper cisterns (see figure 9.6). It causes the float valve to open and shut in rapid succession causing water hammer, or sometimes a noise similar to that of an electric motor. The cure is to stabilise the cistern wall with a strap around it, and to fix the supply pipe securely to prevent it and the connected float valve from moving.

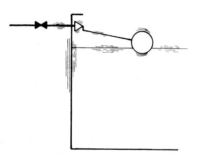

Vibrating cistern wall will move float body and create noise.

Figure 9.6   **Vibration of cistern wall**

Float valve oscillation

This causes a regular series of loud banging noises which occur when the valve is almost closed. This type of noise can be most disturbing to occupiers of buildings, and could well cause damage to pipes and fittings. There are a number of reasons for it, and BS 6700 suggests that the commonest cause is the formation of waves on the surface of the cistern water. I do not agree entirely, in that I believe the waves are a result of the oscillation and not the cause. However, the problem is there whichever is the case!

For the float valve to operate correctly (see figure 9.7), its closing force (float buoyancy) must overcome the incoming force of the water pressure. If the incoming pressure and the buoyancy force nearly balance, the conditions for oscillation are ideal. A shock wave against the washer will cause it to open slightly. The float is depressed into the water and

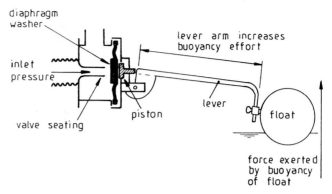

Consider:
(1) size of orifice;
(2) length of lever arm;
(3) diameter of float.

As the cistern water level rises, the closing force exerted on the piston by the float and lever arm must close the diaphragm washer against the incoming force of the inlet pressure.

Figure 9.7   Prevention of float valve oscillation

subsequently bounces back to reclose the valve abruptly, creating a further wave which in turn rebounds to open the valve again; thus a cycle is established.

BS 6700 suggests fitting a damping plate to the float or the lever arm, which will 'dampen down' the effects of the shock waves, or alternatively fitting baffle vanes in the cistern to prevent surface waves affecting the float. In practice these often seem a little 'Heath Robinson'.

Another way of improving flotation is simply to increase the lever arm length, thus achieving better leverage, or to fit a larger float. One could also reduce the size of the orifice seating. Although this might lead to more flow noise, it may be preferable to water hammer.

Tap washer oscillation

This is usually associated with worn, split or soft washers, which vibrate violently as water passes them (see figure 9.8). The noise can be extremely loud but the cure is simple – change the washer!

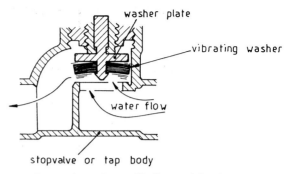

Worn, split or soft washers will vibrate violently.

Figure 9.8   **Tap and valve washer oscillation**

9.3   Other noises

Pumps

These will not cause excessive noise if well designed, unless flow exceeds the pump rating, or the static pressure is insufficient. Noise transmission from pumps can be reduced by:

- using rubber hose isolators between pump and pipework;
- isolating pipework from building structures with resilient inserts fitted in brackets;
- the use of a hydraulic-acoustic filter tuned to the unwanted frequency.

Splashing noises

These occur when water strikes the water surface in cisterns. Silencer tubes screwed into float valve outlets are now prohibited because of the risk of backsiphonage. Some float valves are fitted with collapsible silencer tubes, which are permitted. Others have outlets designed to reduce splashing noise.

Many taps are produced with flow correctors or aerators, which help to reduce water impact noise and splashing into sinks and other appliances. Metal sinks of pressed stainless steel increase splashing sounds, and some treatment in addition to that provided by the manufacturers may be worth considering.

Thermal movement

In hot pipes this causes creaks, squeaks and more impulsive sounds. The use of resilient pipe clips, brackets or pads between pipes and fittings will provide sufficient flexibility to cope with expansion and contraction. Long straight runs are a particular problem for which expansion joints or loops should be considered.

Air or vapour bubbles

These commonly cause flow noise. As a result of poor design and operation of hot water systems, bubbles are formed in hot water cylinders or heaters. Systems should be designed to avoid general or localised boiling and to allow removal of air when filling.

Gases

These are formed by corrosive action in primary circuits and can also cause noise when pumped around systems.

9.4 Noise transmission and reduction

Noise is transmitted to the listener from its source in a number of ways, including direct airborne transmission, through building structures (see figure 9.9), along pipes, and through the water contained in the pipe.

In metal pipes noise is transmitted with little loss, but plastics can reduce noise transmission depending on the pipe material and its thickness. For lengths of between 5 m and 20 m, the reduction is approximately 1 dB/m to 2.5 dB/m.

The insertion of metal-bellows vibration isolators can reduce transmissions by 5 dB to 15 dB, whilst reinforced rubber hose isolators can be even better.

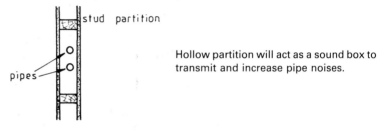

(a) Solid walls

Small flexible pipe
  ○ little noise
Large rigid pipe
  ○ vibration transmitted to wall

Brick will not transmit appreciable sound from small flexible pipes, but larger, more rigid pipes such as steel will induce some vibration.

Hollow partition will act as a sound box to transmit and increase pipe noises.

Lightweight structures vibrate readily, transmitting and accentuating sound from pipes. To prevent this, use lightweight flexible pipes (copper or plastics) and flexible vibration isolating clips.

(b) Hollow walls

Figure 9.9   Structure-borne noises

Resilient mountings can help isolate a storage cistern from its supporting structure, preventing the transmission of noise through ceilings into habitable rooms.

Chapter 10
Accessibility of pipes and water fittings

For many years water byelaws have stated that all pipes and fittings must be readily accessible for inspection, repair and renewal. The new Model is now more specific in that it says:

'...no pipe or other water fitting shall be embedded in any wall or solid floor, or installed in or below a solid floor, or under a suspended floor at ground level...' (Byelaw 58, subject to some exceptions).

While the new byelaws are, in essence, saying the same as the old ones did, the inference is that accessibility of fittings will be more strictly enforced.

However, the degree of accessibility will depend largely upon:

- how strictly the local byelaws are enforced by the water undertaker;
- the personal opinion of the designer, installer and building engineer;
- cost considerations of installation and maintenance of accessible installations using ducts, chases, access panels, etc;
- the consequences of leakages from inaccessible parts of the pipework;
- the reliability of joints, resistance of pipes and joints to internal and external corrosion and flexibility of pipes when being passed through chased ducts or sleeves;
- the importance of aesthetic considerations, on the one hand where pipes are surface mounted and, on the other, the consequences of breaking into, repairing and otherwise spoiling expensive decorations and floor finishings.

10.1 Pipes passing through walls and floors

No pipe may be installed in the cavity of an external cavity wall, other than where it has to pass through from one side to the other (see figures 10.1 and 10.2).

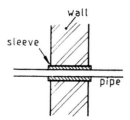

Pipes passing through walls or floors should be in a sleeve to permit movement relative to wall or floor.

Sleeves intended to carry a water pipe should not contain any other pipe or cable.

Figure 10.1   **Pipe sleeved through wall**

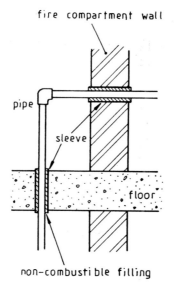

Sleeves should:
- ○ permit ready removal and replacement of pipes;
- ○ be strong enough to resist any external loading exerted by the wall or floor;
- ○ be sealed with fire resistant material that will accommodate thermal movement.

Figure 10.2   **Pipe sleeved through wall and floor**

10.2 Pipes entering buildings

An example of pipes entering building is given in figure 10.3.

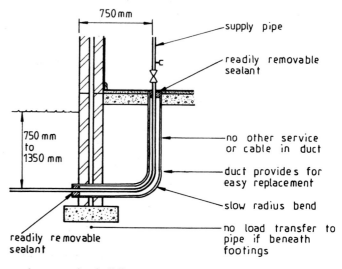

Figure 10.3   **Access to pipes entering buildings**

10.3   Pipes in walls, floors and ceilings

No pipe should be buried in any wall or floor, or under any floor, unless arranged as shown in figure 10.4.

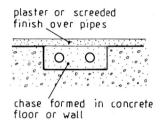

Closed circuit pipes may be screeded or plastered over provided they are in a proper chase, and can be readily exposed for repair or replacement

(a) Pipes plastered or screeded over

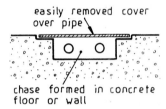

Pipes other than closed circuits must be in a duct that is reasonably easy to expose for examination, repair or replacement, without causing any structural damage.

(b) Pipes under removable cover

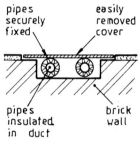

Pipe to be insulated if ducted in an outside wall.

(c) Pipes in external walls

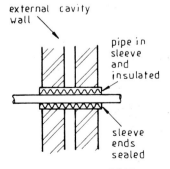

Sleeves passing through cavity walls must pass right through the cavity and the pipe must be insulated. Other than this arrangement no pipes should be laid in a cavity.

(d) Pipes passing through external walls

Figure 10.4   **Pipes in chases and ducts**

continued

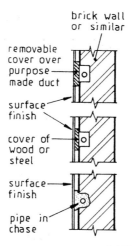

Only permitted if no joints are enclosed and the pipes can be readily withdrawn for examination.

Only permitted if the pipe is an element of a closed circuit for underfloor heating.

(e) Pipes passing through internal walls

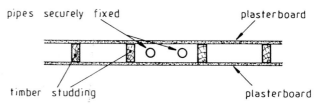

Pipes may be enclosed in internal walls of the timber studded type.

(f) Pipes in stud partition

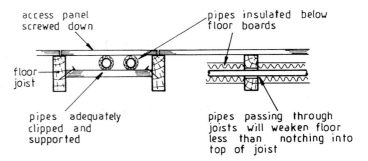

Pipes under suspended floors should be avoided. Where unavoidable they must be insulated and access panels formed in the floor for examination and repair.

Access panels should not affect the stability of floor.

Boards removable at intervals of not more than 2 m and at every joint for inspection of whole length of pipe.

(g) Pipes under timber floors

Figure 10.4 **Pipes in chases and ducts**
continued

continued

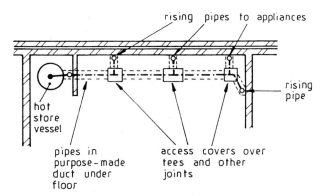

Where pipes are enclosed in a solid floor, access covers should be provided at tees and joints.

(h) Typical ducting arrangements

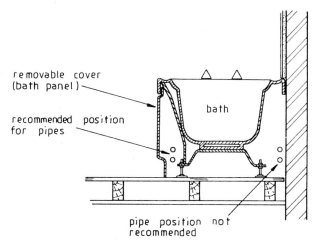

Although BS 6700 recommends that pipes are positioned in front of a bath, the back position gives better fixing for pipes and easier access for bath replacement.

(i) Pipes concealed under baths

Figure 10.4   **Pipes in chases and ducts**
continued

10.4 Stopvalves below ground

Stopvalves on underground pipes should be enclosed within a pipe guard or chamber with surface box to provide access for shutting off with a metal stopvalve key (see figure 10.5).

Stopvalves above ground should be positioned so as to be readily accessible for examination, maintenance and operation.

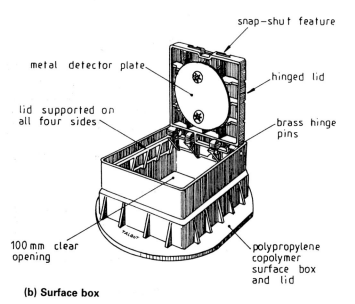

100 mm

stopvalve key

surface box of cast iron or suitable plastic

hinged lid

ground level

surface box

pipe guard

clear space above and around pipe

base

750 mm to 1350 mm

150 mm pipe guard or chamber (drain pipe)

crutch key to fit over stopvalve crutch

brick on edge to form firm base

(a) Chamber construction

Surface box to be suitable for relevant traffic loading, e.g. heavy or light grade.

Stopvalves and pipes more than 1350 mm deep will be deemed to be inaccessible.

snap-shut feature

metal detector plate

hinged lid

lid supported on all four sides

brass hinge pins

100 mm clear opening

polypropylene copolymer surface box and lid

(b) Surface box

Figure 10.5   Access to below ground stopvalves

10.5 Water storage cisterns

These should be positioned so that they are easy to clean and maintain (see figure 10.6).

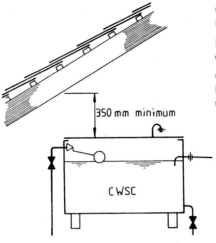

Clear space required all round for inspection of cistern and pipes.

Internal access must be provided for cleaning and maintenance.

Combination units and cisterns in cupboards require 225 mm minimum space.

350 mm minimum

CWSC

(a) Small cisterns

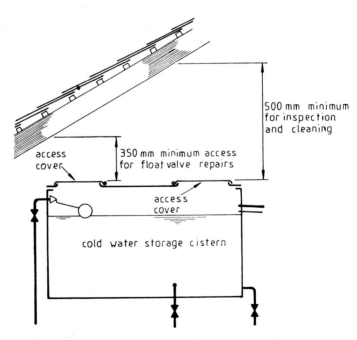

500 mm minimum for inspection and cleaning

access cover

350 mm minimum access for float valve repairs

access cover

cold water storage cistern

Clear space required all round for inspection of cistern and pipes.

(b) Large cisterns

Figure 10.6   Access to cisterns

Chapter 11
Installation of pipework

Pipes and fittings need to be chosen and installed to suit their purpose and the conditions in which they are to be situated.

Pipes should be laid or fixed:

- to avoid frost, mechanical damage or corrosion;
- so that they do not leak, cause undue noise or permit any contamination of water contained in them;
- to comply with the requirements of relevant British Standards or be approved under the UK Water Fittings Byelaws Scheme.

When jointing pipes take care to ensure that:

- joints are mechanically sound;
- pipes are cut squarely and all burrs removed;
- joints comply with British Standard or Water Research Centre requirements;
- when applying heat the risk of fire is eliminated, and the operator does not breathe any harmful fumes given off from soldering or welding processes;
- only approved jointing materials or compounds are used;
- joints, clips and fittings are compatible with the pipe material and will not cause corrosion.

11.1   Steel pipes

There are three grades of low carbon steel pipes to BS 1387: heavy, medium and light, each of which is obtainable with or without galvanization inside and out. For hot and cold services, only galvanized pipes are permitted, as follows:

Heavy – (identified by a red band painted around the pipe near its ends) for use underground where, in addition to galvanization, other forms of protection should be used to guard against exterior corrosion, e.g. a bituminous coating. Exposed threads should be painted.

Medium – (identified by a blue band) for general use above ground only.

Light – (identified by a brown band) permitted on some fire fighting installations.

See examples in figures 11.1–11.3 and tables 11.1 and 11.2.

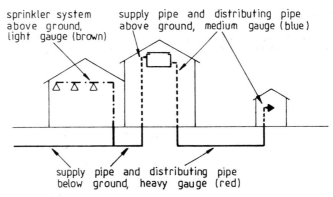

sprinkler system
above ground,
light gauge (brown)

supply pipe and distributing pipe
above ground, medium gauge (blue)

supply pipe and distributing pipe
below ground, heavy gauge (red)

Control valves not shown.

Figure 11.1   Grades of steel pipe and their uses

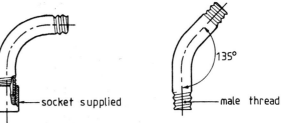

socket supplied with pipe   thread to BS 21

(a) Pipe – standard length 6.4 m

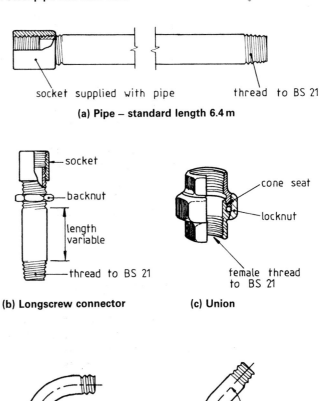

socket

backnut

length
variable

thread to BS 21

cone seat

locknut

female thread
to BS 21

(b) Longscrew connector   **(c) Union**

135°

socket supplied   male thread

(d) 90° pipe bend – male ends   **(e) Obtuse pipe bend – male ends**

Figure 11.2   Selection of fittings for steel tubes to BS 1387

continued

(f) 90° elbow female ends

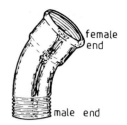

(g) Obtuse pipe bend – male and female

(h) Tee 90° – equal

Numbers indicate method of specifying outlets:
- state 'in line' outlets first;
- state larger end first.

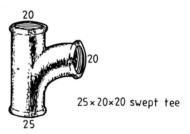

25 × 20 × 20 swept tee

(i) Swept or pitched tee

Figure 11.2 continued   Selection of fittings for steel tubes to BS 1387

Table 11.1   **Thread engagement lengths for steel pipe to BS 1387**

Nominal size of pipe inside diameter mm	Thread length mm
15	13
20	15
25	17
32	19
40	19
50	24
80	30
100	36
150	40

Table 11.2   **Maximum spacing of fixings for internal steel piping**

Nominal size of pipe inside diameter mm	Spacing on horizontal run m	Spacing on vertical run m
15	1.8	2.4
20	2.4	3.0
25	2.4	3.0
32	2.7	3.0
40	3.0	3.6
50	3.0	3.6
80	3.6	4.5
100	3.9	4.5
150	4.5	5.4

(a) Saddle clip of galvanized steel   **(b) Screw-on clip of galvanized cast iron**

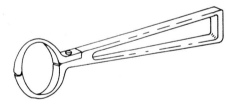

(c) Clip of galvanized cast iron for building in

Figure 11.3   **Pipe fixings for steel tubes**

11.2   Copper pipes

Copper tube to BS 2871 is made in four grades:

Table W – small diameter tubes for microbore heating systems, fully annealed, in coils of 10 m to 200 m length depending on the diameter;

Table X – for normal use above ground, half hard, standard length 6 m;

Table Y – for underground use, fully annealed, produced in coils 20 m long;

Table Z – thin walled, hard drawn, suitable for general use inside buildings, must not be bent or used below ground, standard length 6 m.

Copper tube is also available with a polyethylene coating to add protection from external corrosion. Joints would need wrapping with a suitable adhesive tape after testing. In some areas this may be necessary to prevent corrosion to underground pipes. The polyethylene coating should be coloured blue when the tube is to be used underground. See figures 11.4–11.6 and table 11.3.

Table 11.3   **Maximum spacing of fixings for internal copper and stainless steel piping**

Type of piping	Nominal size of pipe, outside diameter mm	Spacing on horizontal run m	Spacing on vertical run m
Copper (light gauge) and stainless steel complying with BS 2871: Part 1 or BS 4127: Part 2	15	1.2	1.8
	22	1.8	2.4
	28	1.8	2.4
	35	2.4	3.0
	42	2.4	3.0
	54	2.7	3.0
	76	3.0	3.6
	108	3.0	3.6
	133	3.0	3.6
	159	3.6	4.2

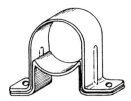

(a) Two piece spacing clip

(b) Single piece spacing clip

(c) Saddle clip

Figure 11.4   **Copper fixing clips for small diameter copper tubes**

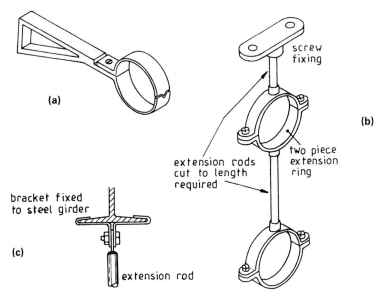

(a)

(b)

screw
fixing

extension rods
cut to length
required

two piece
extension
ring

bracket fixed
to steel girder

(c)

extension rod

(a) **Built-in clip**

(b) **Two piece pipe ring with extension rod and backplate**

(c) **Alternative fixing for pipe ring**

Figure 11.5   **Brass fixing clips for copper tubes**

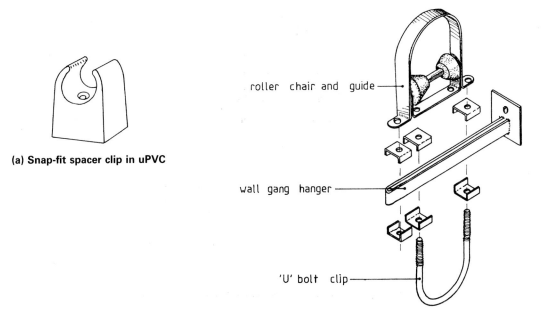

(a) **Snap-fit spacer clip in uPVC**

roller chair and guide

wall gang hanger

'U' bolt clip

(b) **Fixings for large diameter pipes which permit expansion to
take place**

Figure 11.6   **Other fixings for copper tubes**

Jointing methods for copper tubes

Compression fittings – type A (non-manipulative)

These are the most usual type of compression fitting for use above ground. They must not be used below ground.

Figures 11.7–11.12 show various methods for jointing copper tubes.

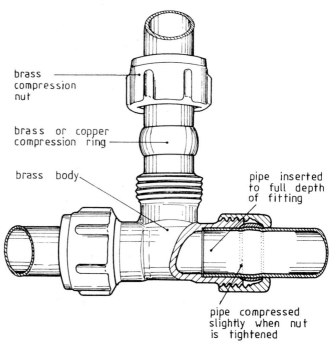

brass compression nut

brass or copper compression ring

brass body

pipe inserted to full depth of fitting

pipe compressed slightly when nut is tightened

For use above ground only.

Also suitable for stainless steel tube.

Do not overtighten brass backnut.

Jointing compounds not required.

(a) Assembly of compression fittings – type A (non-manipulative)

Figure 11.7   Compression fittings – type A

continued

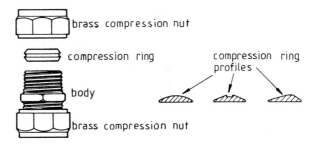

brass compression nut

compression ring

body

brass compression nut

compression ring profiles

Rings from one fitting are not compatible with fittings from another manufacturer.

(b) Straight coupling

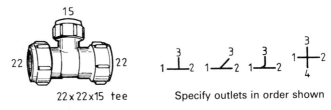

22 × 22 × 15 tee

Specify outlets in order shown

(c) Tee with reduced branch

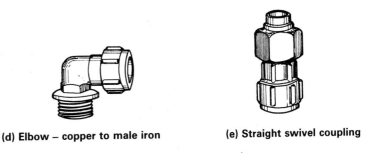

(d) Elbow – copper to male iron

(e) Straight swivel coupling

Supplied with one fibre washer

Figure 11.7 continued **Compression fittings – type A**

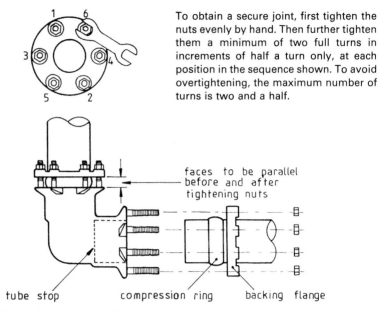

To obtain a secure joint, first tighten the nuts evenly by hand. Then further tighten them a minimum of two full turns in increments of half a turn only, at each position in the sequence shown. To avoid overtightening, the maximum number of turns is two and a half.

faces to be parallel before and after tightening nuts

tube stop   compression ring   backing flange

Figure 11.8   Large diameter compression fitting – type A

Compression fittings – type B (manipulative)

These require the flaring of pipe ends and give extra security against joints pulling apart in extreme conditions. This is especially important underground, where any resulting leakage could well go undetected. See also figures 11.9 and 11.10.

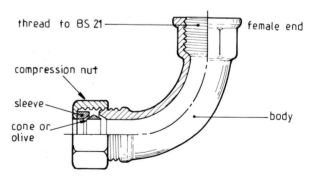

thread to BS 21 ——— female end

compression nut

sleeve

cone or olive

body

Figure 11.9   Type B compression fitting for copper tube – female iron to copper bend

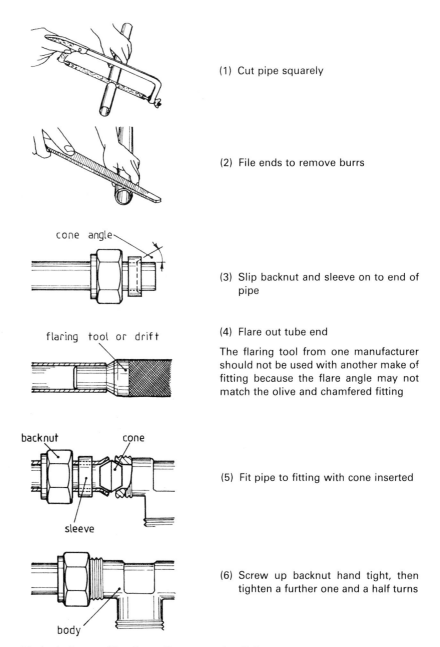

(1) Cut pipe squarely

(2) File ends to remove burrs

(3) Slip backnut and sleeve on to end of pipe

(4) Flare out tube end

The flaring tool from one manufacturer should not be used with another make of fitting because the flare angle may not match the olive and chamfered fitting

(5) Fit pipe to fitting with cone inserted

(6) Screw up backnut hand tight, then tighten a further one and a half turns

Figure 11.10   Method of assembly of type B compression fittings

Fittings have traditionally been made of duplex brass for general use and gunmetal for use below ground, or where dezincification might be a problem. Gunmetal has now been largely replaced by a new dezincification-resistant brass known as DZR and fittings manufactured from DZR will be marked with the following symbol:

Capillary fittings

These are made in two types, as illustrated in figure 11.11. They are suitable for use both above and below ground and are made in a range of copper or brass materials.

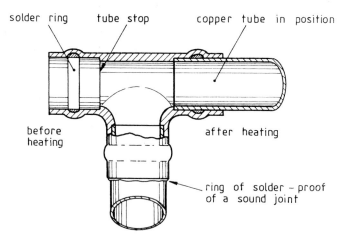

Extra solder should not be added as there is already sufficient within the fitting.

(a) Integral solder ring type

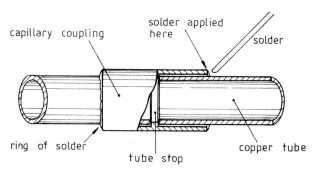

Because the end feed fitting contains no solder, this must be applied to the joint as it is heated.

(b) End feed type

Figure 11.11   Capillary soldered joints

The soldering method is as follows:

(1) Cut tube end square and remove burrs.
(2) Thoroughly clean inside of fitting and outside of pipe with steel wool or sand paper (not emery board).
(3) Apply suitable flux to prevent oxidation and assist cleaning, (just a thin smear to both surfaces).
(4) Assemble joint.
(5) Apply heat with blow torch or electrical tongs.
(6) Add solder to joint area (end feed type only) and continue to apply heat until solder flows around and into joint. For solder ring fittings apply heat until solder appears and forms a ring around the end of the fitting (no extra solder needed).
(7) Leave joint undisturbed to cool.
(8) Remove flux residues or they may corrode the pipe.

Solders have traditionally been lead/tin alloys, the use of which is prohibited by water byelaws in favour of lead-free solders, e.g. tin/silver alloy.

Capillary fittings for copper tube may also be jointed by brazing or silver soldering. This requires much higher temperatures (600°C to 800°C) than soft soldering (180°C to 200°C). Brazed joints are much stronger than soft soldered ones, but much more expensive to install.

Braze (bronze) welded joints (see figure 11.12) are very strong but are even more costly.

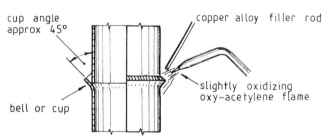

Figure 11.12   **Braze welded bell butt joint**

11.3 Stainless steel pipes

Stainless steel pipe to BS 4127 is similar to copper in use. Its external dimensions are the same as copper. The fittings used are generally those made for copper tubes to BS 864: Part 2.

However, stainless steel pipe is much more rigid than copper and therefore needs a little more accuracy in bending. Bending is done using the copper pipe bending machine.

Pipe diameters range from 15 mm to 159 mm. Spacings for fixings are given in table 11.3.

Jointing methods are similar to those used for copper tubes:

(1) compression fittings of copper alloy or stainless steel;
(2) capillary soldered joints of copper alloy or stainless steel. For soldered joints use phosphoric-acid based flux;
(3) silver soldering and brazing;
(4) anaerobic adhesive bonding (up to 80°C). See figure 11.13.

Note For methods (1) to (3) see jointing methods for copper tube. See also figure 11.10.

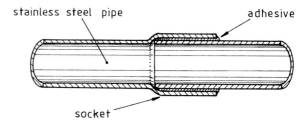

stainless steel pipe   adhesive

socket

Method for anaerobic adhesive bonding
(1) Check fit between tube and fitting.
(2) Ensure bond areas are grease-free using solvent degreaser.
(3) Abrade bonding surfaces with medium emery cloth (80 grit).
(4) Apply ring of adhesive to leading edge of pipe and slip pipe into fitting.
(5) Allow:
    ○ one minute for curing,
    ○ one hour to withstand static pressure,
    ○ one day for full strength.

Figure 11.13   Anaerobic adhesive bonded joint for stainless steel tubes

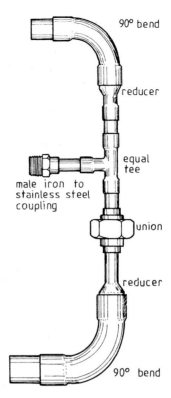

90° bend

reducer

equal tee

male iron to
stainless steel
coupling

union

reducer

90° bend

Figure 11.14 **Selection of joints for stainless steel pipes**

11.4 Polyethylene pipes

Polyethylene has proved to be an excellent material for cold water installations, particularly when used below ground. It is corrosion resistant, easy to lay and simple to joint. It can be obtained in long lengths permitting supply pipes to be laid with the minimum of joints. Its flexibility allows it to be bent around obstacles and threaded through ducts into buildings. It must not be used for hot water installations.

When laying polyethylene pipe below ground, it is advisable to bed and cover the pipe with a selected soil, or granular material such as pea shingle, to prevent damage from stones and flints, and to avoid deformation of the pipe when backfilling the trench.

In the past polyethylene has been known to be permeable to gas, and is liable to damage by contact with oils or oil-based products.

The British Standards Institution is, at the time of writing, completely reviewing its specifications for polyethylene pipes. New specifications are being produced to replace the long-established ones. The reasons for the review are as follows:

(1) Metrication, which began some 20 years ago, has been slow to catch up with the plastics pipes. This is now being rectified.

(2) Simplification of the range of pipe sizes and pressure ratings. There will be medium density polyethylene pipes in one pressure range only, instead of both high density and low density materials each produced for three pressure ranges.

(3) Identification of pipes below ground. Polyethylene pipe for use below ground is to be coloured blue for instant identification and to prevent confusion between services.

Table 11.4 gives information on the various polyethylene pipes including those for which the standards are obsolescent or have been withdrawn.

Table 11.4   Types and grades of polyethylene pipes

Type of piping	Grade	Maximum working pressure bar	Type of piping	Nominal size, outside diameter mm	Approximate bore mm	Maximum working pressure bar
Low density polyethylene (Type 32) to BS 1972 and high density polyethylene (Type 50) to BS 3284	Class B (light)	6	Blue medium density polyethylene to BS 6572 for below ground use only and black medium density polyethylene to BS 6730 for above ground use only	20	15	12
	Class C (medium)	9		25	20	
				32	25	
	Class D (heavy)	12		50	40	
				63	50	
Preferred lengths: Straight pipes 6 m, 9 m, 12 m. Coils 50 m, 100 m, 150 m.						

Notes
(1)   BS 1972 is obsolescent and BS 3284 is withdrawn.
(2)   A further British Standard is under preparation for blue polyethylene pipes in sizes from 90 mm to 310 mm. Currently these sizes are available with Water Research Centre approval.

Fixing distances for polyethylene pipes

Because polyethelene is a flexible material, pipes above ground should be supported continuously, or at least be fixed to the spacing requirements shown in table 11.5.

Table 11.5 **Maximum spacing of fixings for internal polyethylene pipes**

Type of piping	Nominal size of pipe	Spacing on horizontal run m	Spacing on vertical run m
Low density polyethylene to BS 1972 and medium density polyethylene to BS 6730 (see note)	$\frac{3}{8}$	0.30	0.60
	$\frac{1}{2}$	0.40	0.80
	$\frac{3}{4}$	0.40	0.80
	1	0.40	0.80
	$1\frac{1}{4}$	0.45	0.90
	$1\frac{1}{2}$	0.45	0.90
	2	0.55	1.10
	$2\frac{1}{2}$	0.55	1.10
	3	0.60	1.20
	4	0.70	1.40
High density polyethylene to BS 3284 Type 50	$\frac{3}{8}$	0.45	0.90
	$\frac{1}{2}$	0.60	1.20
	$\frac{3}{4}$	0.60	1.20
	1	0.60	1.20
	$1\frac{1}{4}$	0.70	1.40
	$1\frac{1}{2}$	0.70	1.40
	2	0.85	1.70
	$2\frac{1}{2}$	0.85	1.70
	3	0.90	1.80
	4	1.10	2.20
	6	1.30	2.60

Note   There are no spacings available for medium density polyethylene pipes, therefore use the spacings for low density polyethylene pipes.

Jointing methods for polyethylene pipes

Compression fittings

These are shown in figures 11.15–11.17.

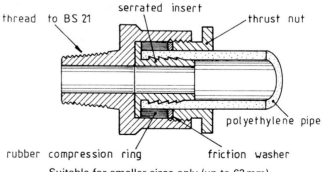

thread to BS 21   serrated insert   thrust nut

rubber compression ring   friction washer

polyethylene pipe

Suitable for smaller sizes only (up to 63 mm).

Figure 11.15   **Compression fitting for polyethylene pipes**

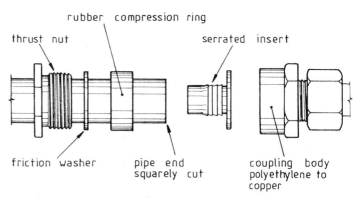

This illustration and the instructions for making the joint are for a compression joint manufactured by Talbot. Similar fittings are available from several other manufacturers.

(1) Assemble parts in sequence shown.

(2) Using a hide mallet or similar, knock insert into pipe end until its flange touches pipe face.

(3) Push rubber compression ring and friction washer against flange and locate assembly in fitting body.

(4) Screw in thrust nut until hand tight, then tighten fully with spanner (one and a half to two turns).

Figure 11.16   Making the compression joint

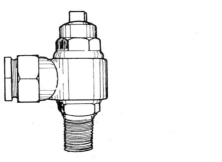
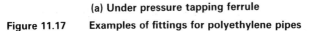
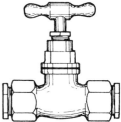

(a) Under pressure tapping ferrule

(b) Stopvalve

Figure 11.17   Examples of fittings for polyethylene pipes

Push-fit joints

These are shown in figures 11.18–11.21.

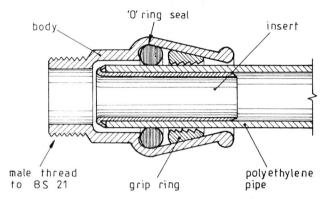

Figure 11.18   **Push-fit joint for polyethylene pipe**

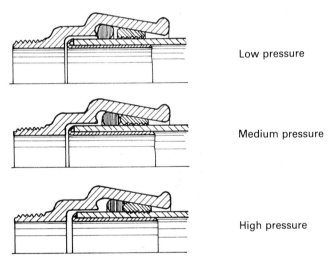

Low pressure

Medium pressure

High pressure

As water pressure increases the components tighten to create a pressure seal.

Figure 11.19   **How the push-fit joint works**

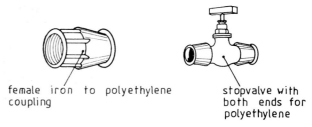

female iron to polyethylene coupling

stopvalve with both ends for polyethylene

These fittings are available in a variety of sizes from 20 mm to 180 mm

Figure 11.20   **Examples of small diameter fittings**

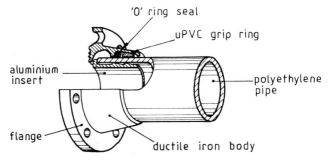

'O' ring seal

uPVC grip ring

aluminium insert

polyethylene pipe

flange

ductile iron body

Figure 11.21   **Example of large diameter fitting – flange adaptor**

Thermal fusion joints

These joints, illustrated in figure 11.22, are suitable for large diameter polyethylene (PE) pipes, but are not popular for small diameters because other methods of jointing are so much easier. They are also suitable for polypropylene (PP) pipes. In addition to the jointing method shown in figure 11.23, polyethylene pipes can be jointed by 'electrofusion' and butt fusion.

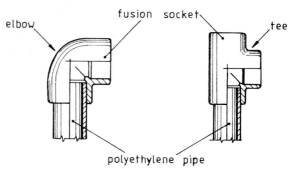

elbow

fusion socket

tee

polyethylene pipe

Figure 11.22   **Thermal fusion joints**

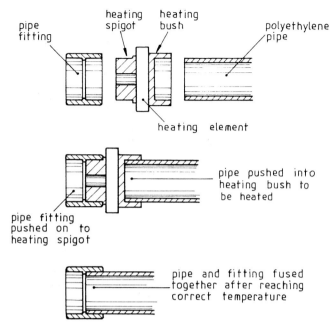

Also suitable for polypropylene pipe.

Pipes should only be joined with compatible fittings, i.e. pipe and fittings in polyethylene or pipe and fittings in polypropylene.

Correct heat is essential.

Follow manufacturer's instructions for jointing method.

Figure 11.23   Making thermal fusion joints

11.5   uPVC pipes

uPVC (unplasticized polyvinylchloride to BS 3505) is an excellent material for cold water pipes for temperatures up to 20°C, beyond which its mechanical properties are reduced. Available in three pressure ranges and a variety of diameters, from 10 mm to 600 mm (see table 11.6), it is used in great quantities for water mains but is not so popular for smaller services. It is equally suited to all installations above and below ground, for cold water only. A similar material, cPVC (chlorinated polyvinylchloride), is suitable for use with hot water.

Table 11.6 **Sizes and pressure ratings of uPVC pipe to BS 3505**

Pressure class	Pressure rating at 20°C bar	Range of nominal sizes, inside diameter	
		inches	mm
C	9	2 to 24	50 to 600
D	12	$1\frac{1}{4}$ to 18	32 to 450
E	15	$\frac{3}{8}$ to 16	10 to 400

Fixings for uPVC pipes

Examples of fixings and their spacing are shown in figure 11.24 and table 11.7.

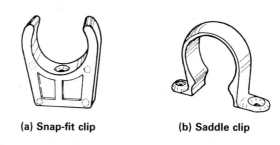

(a) Snap-fit clip   **(b) Saddle clip**

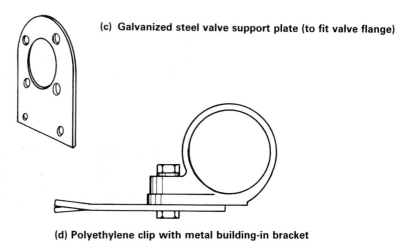

(c) Galvanized steel valve support plate (to fit valve flange)

(d) Polyethylene clip with metal building-in bracket

Figure 11.24   **Fixings for uPVC pipes**

Table 11.7   **Maximum spacing of fixings for internal uPVC and cPVC pipes**

Type of piping	Nominal size of pipe inches		mm	Spacing on horizontal run m	Spacing on vertical run m
uPVC to BS 3505	$\frac{3}{8}$		10	0.530	1.06
(Figures are for	$\frac{1}{2}$		13	0.610	1.22
normal ambient	$\frac{3}{4}$		20	0.685	1.37
temperatures	1		32	0.760	1.52
below 20°C. For	$1\frac{1}{4}$		32	0.840	1.68
temperatures	$1\frac{1}{2}$		40	0.915	1.83
above 20°C the	2		50	1.065	2.13
pipe manufacturer	3		75	1.370	2.74
should be	4		100	1.525	3.05
consulted)	6		150	1.830	3.66
cPVC (Based	$\frac{3}{8}$	and	12	0.6	1.2
on average	$\frac{1}{2}$	and	15	0.8	1.6
temperature of	$\frac{3}{4}$	and	22	0.8	1.6
80°C)	1	and	28	0.9	1.8
	$1\frac{1}{4}$	and	32	1.0	2.0

Jointing methods for uPVC pipes

Threaded joints

These should not be used for uPVC pipes to BS 3505. However, this method can be used for pipe to BS 3506 class 7 up to 2 in diameter, providing the pressure does not exceed 9 bar.

Compression joints

See figure 11.25. These are similar to those for polyethylene. BS 6700 suggests that joints must be made of uPVC also, but there is at least one manufacturer who produces approved fittings of brass and gunmetal.

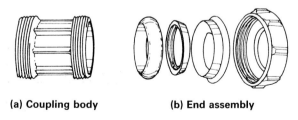

(a) Coupling body   **(b) End assembly**

Similar fitting can be used for polyethylene pipes

Figure 11.25   **Compression joint for uPVC pipes**

Mechanical joints

See figure 11.26 as an example.

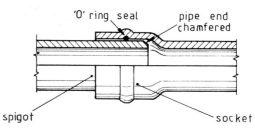

The push-fit joint for uPVC is the simplest and cheapest form of installation.

Care should be taken to ensure that bends and branches are secure and properly anchored before pressure is applied.

Pipe end should be lubricated with suitable bactericidal, non-toxic lubricant to assist insertion into socket.

Figure 11.26   **Mechanical joint for uPVC pipe (push fit type)**

Solvent cement joints

See figures 11.27 and 11.28 as examples.

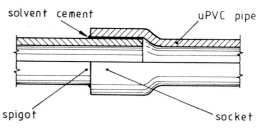

Jointing method:
(1) Cut pipe end square and remove internal burrs.
(2) Slightly chamfer outer edge of pipe (at about 15° to pipe axis).
(3) Roughen joint surfaces, using clean emery cloth or medium glasspaper.
(4) Degrease joint surfaces of pipe and fitting with cleaning fluid, using absorbent paper.
(5) Using a brush, apply an even layer of cement to both fitting and pipe in a lengthwise direction, with a thicker coating on the pipe.
(6) Immediately push the fitting on to the pipe without turning it. Hold for a few seconds, then remove surplus cement.
(7) Leave undisturbed for five minutes, then handle with reasonable care.
(8) Allow 8 hours before applying the full rated pressure, and 24 hours before testing at one and a half times the full rated pressure.
For lower pressure, allow one hour per bar, e.g. 3 bar would require 3 hours drying time.

Figure 11.27   **Solvent cement joint for uPVC pipe**

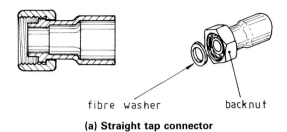

(a) **Straight tap connector**

(b) **Equal tee 90°**

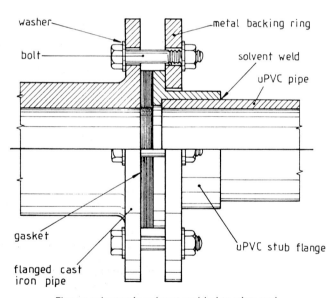

Flange adaptor is solvent welded to pipe end.

(c) **Flanged connections for uPVC**

Figure 11.28   **Examples of solvent cement joints**

11.6   Iron pipes

These are made in three types: vertically cast, spun iron and ductile iron. However, the production of vertically cast and spun iron pipes has now virtually ceased in favour of pipes in ductile iron to BS 4772, which have much improved mechanical properties.

Since all iron pipes are liable to corrosion they are factory treated inside and out. Also, many water authorities may require pipes to be sheathed in a blue polyethylene sleeve to BS 6076 for further protection from aggressive soils, and for identification.

Ductile iron pipes up to 1600 mm in diameter are considered suitable for working pressures of up to 40 bar depending on the size and pressure rating, and should be pressure tested to 5 bar above the expected working pressure.

Fixings for cast iron pipes

See figure 11.29 and table 11.8 for examples of fixings and spacings for cast iron pipes.

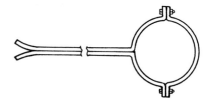

(a) Holderbat build-in type in mild steel

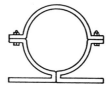

(b) Holderbat screw-to-wall type in mild steel

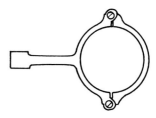

(c) Holderbat hinged build-in type in cast iron

Figure 11.29   Brackets for cast iron pipe

Table 11.8   Maximum spacing of fixings for above ground cast iron pipes

Type of piping	Nominal size of pipe mm	Spacing on horizontal run m	Spacing on vertical run m
Vertically cast or spun iron complying with BS 1211 or BS 2035 and ductile iron complying with BS 4772	51	1.8	1.8
	76	2.7	2.7
	102	2.7	2.7
	152	3.6	3.6

Jointing methods for cast iron pipes

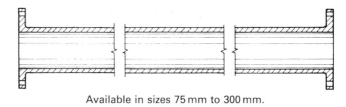

Available in sizes 75 mm to 300 mm.

Figure 11.30   Flanged spun iron and cast iron pipes

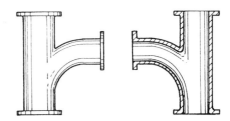

Figure 11.31   Flanged junction

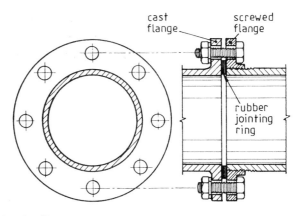

Figure 11.32   Flanged joint detail

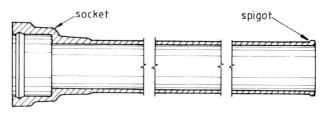

Plain joint for lead.

Figure 11.33   **Cast iron pipe with plain socketed joint**

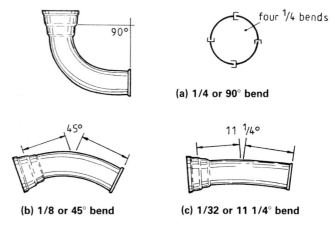

(a) 1/4 or 90° bend

(b) 1/8 or 45° bend

(c) 1/32 or 11 1/4° bend

Figure 11.34   **Plain socketed pipe bends**

Caulking tool chosen to match pipe size and width of joint.

Caulk when cold to finish about 3 mm inside socket face.

Use synthetic yarn that will not promote the growth of bacteria. Yarn should be caulked tightly to approximately one-third depth of the joint, to prevent direct contact between lead and water, and to centralise pipe in socket.

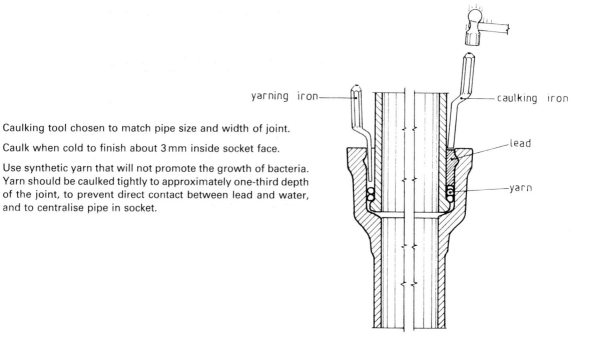

Figure 11.35   **Lead run joint for cast iron pipes**

Ductile iron pipe will use flanged joints or more commonly push-fit 'O' ring type mechanical joints (see figures 11.36 and 11.37), giving good flexibility of movement without loss of joint seal. See also figures 11.30 – 11.35.

Other joints used for cast iron pipes are the Viking Johnson (see figure 11.38) and the bolted gland joint (see figure 11.39).

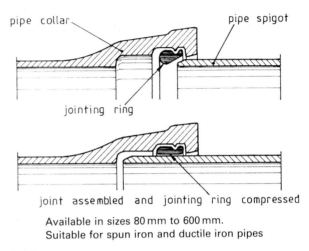

Available in sizes 80 mm to 600 mm.
Suitable for spun iron and ductile iron pipes

Figure 11.36   **'Tyton' slip-fit joint**

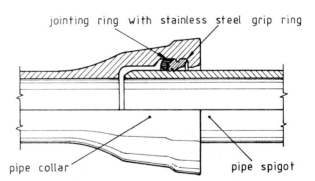

Stainless steel grip ring makes joint resistant to pulling apart.

Figure 11.37   **Slip-fit joint with self-anchoring gasket**

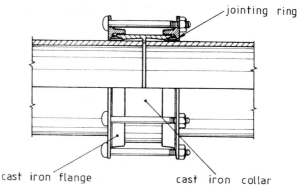

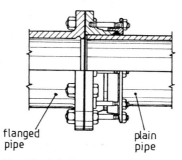

Also suitable for uPVC, steel and asbestos cement pipes.

(a) Straight coupler

Used for jointing plain ended pipes to flanged pipes or fittings.

Each adaptor consists of flanged sleeve, end flange, wedge rubber packing ring, bolts, studs and nuts.

(b) Flange adaptor

Figure 11.38   Viking Johnson joints

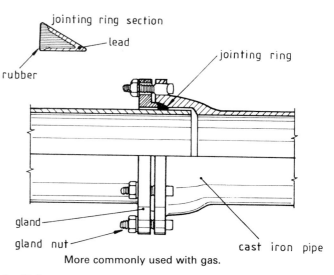

More commonly used with gas.

Figure 11.39   Bolted gland joint

11.7   Asbestos cement pipes

Asbestos cement pipes, in common with many asbestos-based materials, are subject to the requirements of the Asbestos Regulations 1969 and the Health and Safety at Work Act 1974. These regulations require that any dust liberated is restricted to a low level of concentration. Asbestos cement pipes contain only a small percentage of asbestos and are considered safe to handle. Where a limited amount of cutting and turning by hand tools is done in the open air, the dust level is generally below the minimum set in the regulations. However, if any doubt exists, clarification must be sought from the local office of HM Factory Inspectorate, or from the manufacturer of the material.

Asbestos cement pipe is corrosion resistant in most soils but since it is very brittle it is liable to break where soils, such as clays, move because of seasonal loss or gain of moisture.

The use of asbestos cement pipes is restricted to installations below ground. Pipe is available from 50 mm to 900 mm diameter. See table 11.9 and figures 11.40 and 11.41.

Table 11.9   **Class and pressure range of asbestos cement pipes**

Class	Metric colour coding (pipes, joints and rings)	Works test pressure bar	Maximum working pressure bar	Test pressure
15	Green	15	7.5	One and a half to twice the expected working pressure
20	Blue	20	10	
25	Violet	25	12.5	

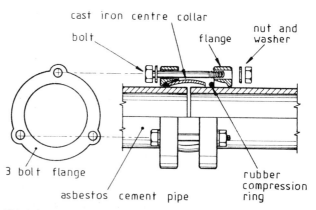

This joint needs careful protection from corrosion otherwise the advantage of the non-corrosive properties of asbestos cement pipes is lost.

Made also for cast iron pipes.

Figure 11.40   Detachable joint made of cast iron

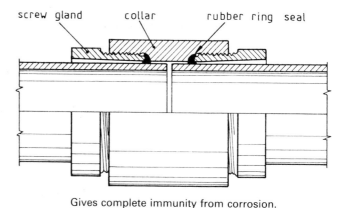

Gives complete immunity from corrosion.

Figure 11.41   All asbestos cement screwed gland joint

11.8   Connections between pipes of different materials

As far as possible connections between different pipe materials should be avoided especially when jointing dissimilar metals, as this may lead to electrolytic corrosion. However, when dissimilar pipes have to be used, for example, in the repair or renewal of part systems or connections to existing pipelines, they may be connected by one of the following three methods:

(1) using 'inert' material between the dissimilar metals (see figure 11.42);

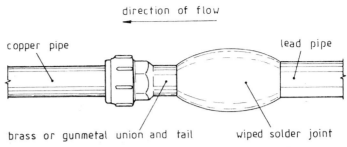

direction of flow

copper pipe

lead pipe

brass or gunmetal union and tail

wiped solder joint

Brass or gunmetal tail used to prevent direct contact between copper pipe and lead pipe. Direct contact is not permitted.

The wiped lead soldered joint is listed in Appendix F of BS 6700 as a means of connecting to existing lead pipes.

An alternative jointing method is to use a 'patent' compression joint.

Figure 11.42 **Use of inert material between connections of dissimilar metals**

(2) ensuring that water flows towards a potentially stronger material in the electrochemical series from a weaker one:

copper ◄──── lead ◄──── iron or steel ◄──── galvanized steel
(uncoated)

direction of water flow

Water flowing in the opposite direction will carry dissolved particles of potentially stronger material which will adversely affect the weaker pipeline;

(3) using a sacrificial anode of a potentially weak material (see figure 11.43).

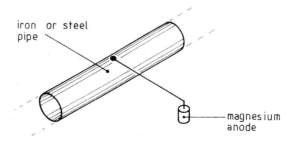

iron or steel pipe

magnesium anode

Anode will corrode and leave the pipe intact.

Sacrificial anodes can be fitted to cisterns, tanks and cylinders.

Figure 11.43 **Sacrificial anode used on a pipeline**

11.9 Connections to cisterns and tanks

The following general instructions should be followed. See also figure 11.45.

(1) Provide proper support for cisterns and tanks to avoid undue stress on connections and deformation of cistern or tank walls.
(2) Use proper tools for hole cutting (see figure 11.44) not flame cutters, not hammer and chisel.
(3) Holes must be truly circular with clean edges.
(4) All debris or filings must be removed from inside the tank or cistern.

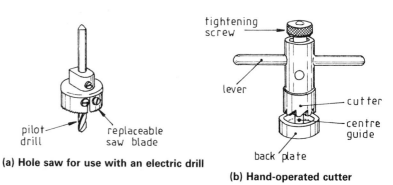

(a) Hole saw for use with an electric drill

(b) Hand-operated cutter

Figure 11.44   Tools for cutting holes

Additional considerations for thermo-plastics cisterns.

(1) Scribing tools should *not* be used to mark position of holes.
(2) Cistern wall should be supported with wooden strut or similar during cutting.
(3) Pipes should be carefully fitted and supported to avoid distortion of cistern or tank.
(4) Corrosion resisting support washers should be used inside and outside the cistern to strengthen the joint area.
(5) Float valves should be fitted through a supporting back plate to stabilize the cistern wall against the thrust of the lever arm.
(6) Linseed oil based sealants must not be used with plastic cisterns or pipes. Use only PTFE sealants.

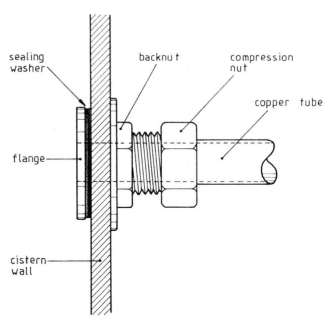

Example shows type A compression joint to copper tube

(a) Copper connection to cistern

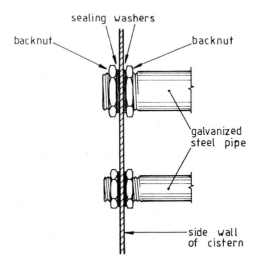

Copper pipe should not be connected to galvanized steel cisterns or tanks.

Pipe sealed with proprietary washers.

(b) Galvanized steel connections to steel cistern

Figure 11.45   Examples of cistern connections.

continued

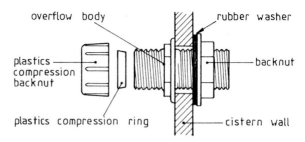

(c) Plastics overflow connection to WC cistern

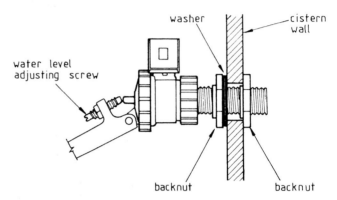

(d) Plastics float valve connection to cistern

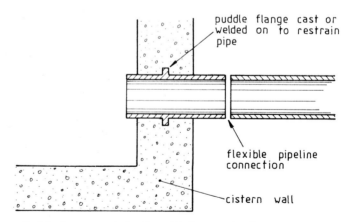

Puddle flange properly aligned before casting into concrete.

Concrete to be compacted to ensure watertight joint.

(e) Connection to concrete tank or cistern

Figure 11.45 **Examples of cistern connections**
continued

11.10 Branch connections for buildings

Service connections to mains are normally made by drilling and tapping the top of the main and screwing in a union ferrule (see figures 11.46 and 11.47).

The ferrule is set on top of the main to avoid drawing sediment into the service pipe.

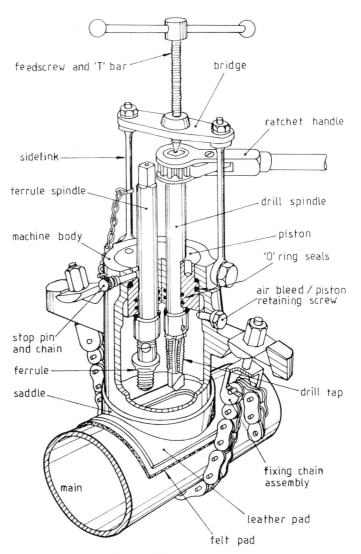

feedscrew and 'T' bar

bridge

ratchet handle

sidelink

ferrule spindle

drill spindle

piston

machine body

'O' ring seals

air bleed / piston retaining screw

stop pin and chain

ferrule

saddle

drill tap

main

fixing chain assembly

leather pad

felt pad

Figure 11.46   Under pressure mains tapping machine

Ferrule screwed directly into top of cast iron water main.

Branch pipe to run parallel to main before taking its final route.

Gooseneck to permit movement in soil without damage to pipe.

Gooseneck bent to ensure that ferrule will not loosen if the service pipe should settle.

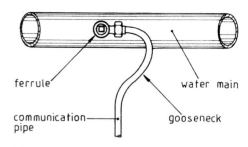

(a) Connection of ferrule to main

Mains pipes of asbestos cement, steel and uPVC are strengthened at the point of connection by the use of a strap saddle.

Strap saddle of cast iron for asbestos cement and steel mains, and of gunmetal for uPVC mains.

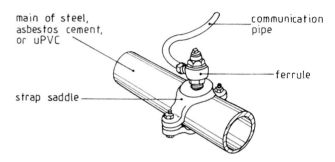

(b) Use of strap saddle for mains connections

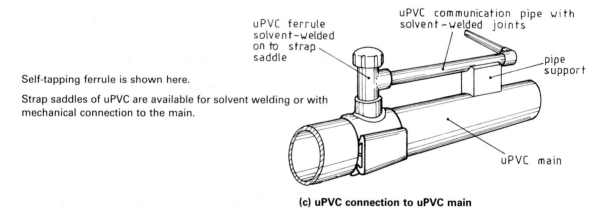

Self-tapping ferrule is shown here.

Strap saddles of uPVC are available for solvent welding or with mechanical connection to the main.

(c) uPVC connection to uPVC main

Figure 11.47   Mains connection

Where rigid pipes are used for the branch service pipe connect them using a short length of suitable flexible pipe.

Depending on the size of the service pipe and the main, ferrules may not always be suitable. For larger connections or smaller mains a more suitable method would be the use of a leadless collar or a tee connection (see figures 11.48 and 11.49, and table 11.10).

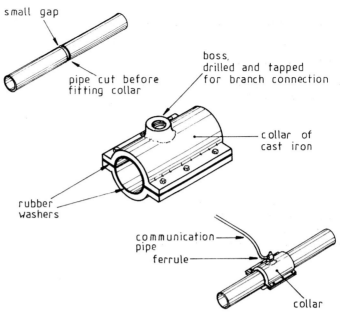

Figure 11.48 **Leadless collar**

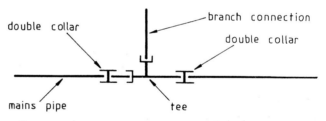

Figure 11.49 **Tee connection**

Table 11.10 **Method of branch pipe connection**

Nominal size of branch pipes		Nominal diameter of main pipe				
mm	inches	80 mm	100 mm	150 mm	200 mm	250 mm and over
15	$\frac{1}{2}$	F	F	F	F	F
22	$\frac{3}{4}$	T	F	F	F	F
25	1	T	T	F	F	F
35	$1\frac{1}{4}$	T	T	T	F	F
42	$1\frac{1}{2}$	T	T	T	F	F
54	2	T	T	T	T	F

F: Ferrule   T: Tee or leadless collar

11.11   Contamination of mains

When cutting into mains to make branch connections, precautions should be taken to avoid contamination.

(1) Sterilize the trench around the branch connection before cutting.
(2) Take care to avoid the entry of soil or water from the trench.
(3) Insert sterilizing tablets into the pipe when making the connection.

11.12   Laying underground pipes

Pipes should be laid on a firm even base, evenly supported throughout their length, and must not rest on their sockets, bricks or other makeshift supports.

As far as possible pipes should be laid in straight lines to permit easy location later. However, copper or plastics pipes should be snaked within the trench to allow for settlement and moisture movement in the soil (see figure 11.50).

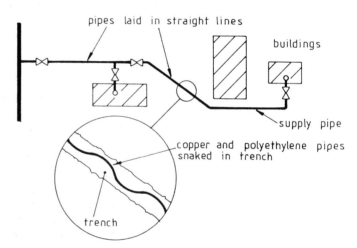

Large diameter uPVC pipes must not be bent within their length. However, some deflection is permissible at flexible joints.

Figure 11.50   **Location of pipelines**

Trench excavations

The bottoms of trenches must be carefully prepared to a firm even surface so that the barrels of the pipes when laid are well bedded down for their whole length (see figure 11.51). Mud, rock projections, boulders, hard spots and local soft spots should be removed and replaced with selected fill material consolidated to the required level. The width of trenches must be sufficient to allow the pipes to be properly bedded, jointed and backfilled.

Where rock is encountered, the trench should be cut at least 150 mm deeper than other ground and made up with well rammed bedding material.

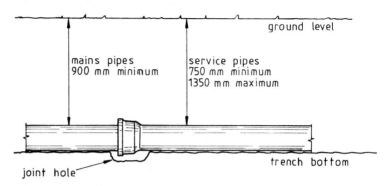

Pipeline to follow contours of ground level whilst ensuring adequate depth of cover.

Pipe to lie flat on trench bottom if soil conditions permit.

Additional excavation needed to leave space for collar.

Figure 11.51   Support of pipes in trench

Pipes must be kept clean and each pipe and fitting must be thoroughly cleansed internally immediately before fitting. The open end should be temporarily plugged until jointing takes place. Particular care must be taken to keep the joints clean. After laying and jointing, the leading end must remain plugged.

Precautions must be taken to prevent the plugged pipes floating if the trench becomes flooded.

Coatings, sheathings or wrappings must be examined for damage, and repaired where necessary, before trenches are backfilled.

When backfilling trenches (see figure 11.52), the pipes must be surrounded by a selected material consolidated to prevent subsequent pipe movement. No large stones or sharp objects should be in contact with the pipes.

uPVC pipes need extra care, especially when being backfilled, otherwise pipes will become distorted and weakened. Granular bedding and surrounds are essential (see figure 11.53).

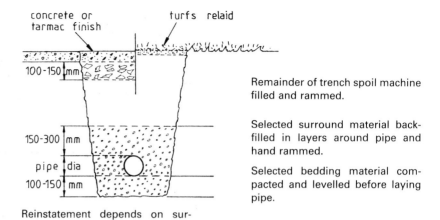

concrete or tarmac finish   turfs relaid

100-150 mm

150-300 mm

pipe dia

100-150 mm

Remainder of trench spoil machine filled and rammed.

Selected surround material back-filled in layers around pipe and hand rammed.

Selected bedding material compacted and levelled before laying pipe.

Reinstatement depends on surrounding surface, e.g. grass verge or road.

Figure 11.52   Trench backfilling

Light compacting by machine

Compact by hand in layers

360° granular surround will give all-round protection and prevent ovalling of pipe.

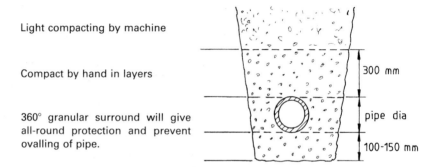

300 mm

pipe dia

100-150 mm

Figure 11.53   Backfilling trenches for uPVC pipes

Restraint of pipes

With most methods of pipe laying and trench backfill for large diameter pipes, joints at changes in direction are liable to move and push apart due to internal thrust pressure (see figure 11.54). To guard against these risks, which will vary according to the internal pressure and how much the pipe direction changes, pipes must be securely anchored at bends, branches and pipe ends. The amount of anchorage required will also depend upon the soil and its bearing capacity. See also tables 11.11–11.13.

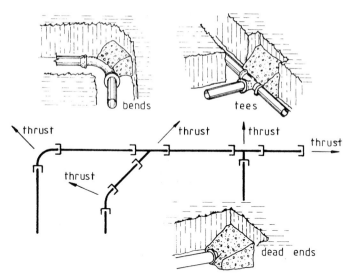

Figure 11.54   **Horizontal thrust on buried mains**

Table 11.11   **Thrust per bar internal pressure**

Nominal internal diameter of pipe	End thrust	Radial thrust on bends of angle			
		90°	45°	$22\frac{1}{2}°$	$11\frac{1}{4}°$
mm	kN	kN			
50	0.38	0.53	0.29	0.15	0.07
75	0.72	1.02	0.55	0.28	0.15
100	1.17	1.66	0.90	0.46	0.24
125	1.76	2.49	1.35	0.69	0.35
150	2.47	3.50	1.89	0.96	0.49
175	3.29	4.66	2.52	1.29	0.65
200	4.24	5.99	3.24	1.66	0.84
225	5.27	7.46	4.04	2.06	1.04
250	6.43	9.09	4.92	2.51	1.26
300	9.38	13.26	7.18	3.66	1.84
350	12.53	17.71	9.59	4.89	2.46

Table 11.12   **Gradient thrusts on buried or exposed mains**

Gradient	Spacing of anchor block		
	m	ft	
1 in 2	5.5	18	
1 in 3	11.0	36	
1 in 4	11.0	36	gradient 1 in 6 or steeper
1 in 5	16.5	54	
1 in 6	22.0	72	

Table 11.13   **Bearing capacity of soils**

Soil type	Safe bearing load kN/m^2
Soft clay	24
Sand	48
Sandstone and gravel	72
Sand and gravel bonded with clay	96
Shale	240

Joints below ground and in other inaccessible places should be kept to a minimum and avoided where possible. It is preferable that pipes should be laid in one length. This applies particularly to underground supply pipes.

Surface boxes must be provided to allow access to valves and hydrants, and must be supported on concrete or brickwork which, after making an allowance for settlement, must not rest on the pipes and transmit loads to them. See figure 11.55.

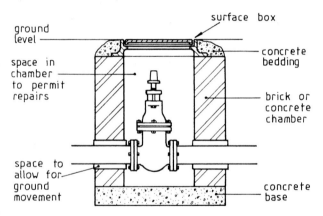

Figure 11.55   **Valve chamber**

Brick or concrete hydrant chambers (see figure 11.56) must be large enough to permit repairs to be carried out to the fittings.

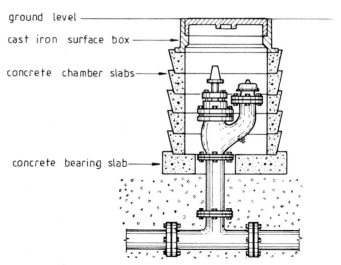

ground level

cast iron surface box

concrete chamber slabs

concrete bearing slab

Concrete chamber slabs provide quick and easy method of building chamber plus easy removal for access for repairs.

Hydrant installations to comply with BS 3251.

Internal dimensions 430 mm × 280 mm.

Figure 11.56 **Typical hydrant arrangement**

An alternative is to provide vertical guard pipes or precast concrete sections to enclose the spindles of valves as shown in figure 11.57.

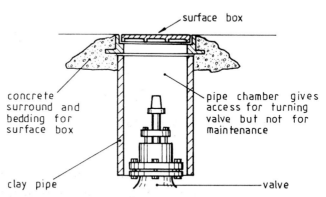

surface box

concrete surround and bedding for surface box

pipe chamber gives access for turning valve but not for maintenance

clay pipe

valve

Commonly used because it is cheap, but valve must be dug out for repairs.

Figure 11.57 **Valve access**

All valves and hydrants should be positioned where they can easily be found and used.

Surface boxes must be positioned and marked to indicate the pipe service, the size of mains, and position and depth below the surface (see figure 11.58).

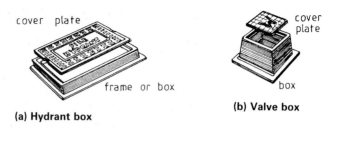

(a) Hydrant box

(b) Valve box

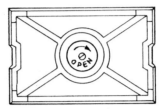

(c) Hydrant cover marking

Cover plates marked to identify type of valve enclosed in chamber.

Hydrant cover must be marked underneath to indicate direction of valve turning.

Figure 11.58   Surface boxes for valves and hydrants

Indicator plates, illustrated in figure 11.59, can be screwed to walls or marker posts, but must be visible.

Drawings should show all pipe runs, valves and hydrants. Working drawings should be amended to show variations from the original design.

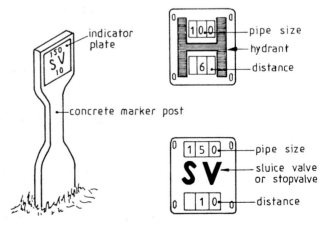

Figure 11.59   Valve and hydrant marker plates

11.13   Pipework in buildings

Fixings and allowance for thermal movement

Allowance must be made for expansion and contraction of pipes (see figure 11.60), especially where pipe runs are long and where temperature changes are considerable (hot distributing pipes) and where the pipe material has a relatively high coefficient of thermal expansion.

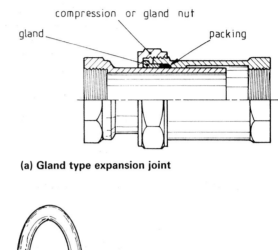

(a) Gland type expansion joint

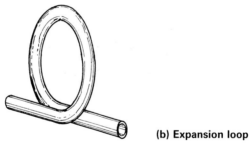

(b) Expansion loop

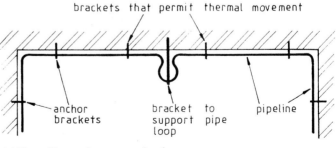

(c) Use of horseshoe expansion loop

Figure 11.60   **Expansion joints**

Fixing insulated piping

Sufficient space should be allowed behind pipes for insulation to be properly installed, as shown in figure 11.61.

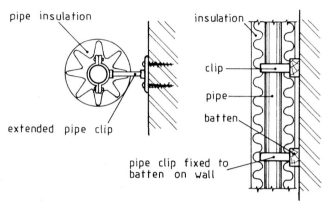

(a) Extended bracket   **(b) Bracket fixed to battens**

Figure 11.61   Fixings for insulated piping

Concealed pipes

These must be housed in properly constructed ducts or chases that provide access for maintenance and inspection. See chapter 10.

Structural members

These must not be weakened by indiscriminate notching and boring. Notches and holes should be as small as practicable. Whenever possible, notches should be U-shaped and formed by parallel cuts to previously bored holes (see figure 11.62). The positions of notches and holes are as important as their sizes, and in timber beams and joists should only be made within the hatched areas shown in figure 11.63.

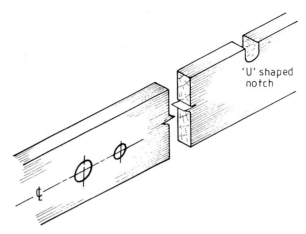

'U' shaped notch

Holes and notches to be as small as practicable but large enough to permit pipes to expand and contract.

To avoid weakening the joist, holes and notches should be restricted to the hatched areas shown in figure 11.63.

Figure 11.62   Pipes passing through structural timbers

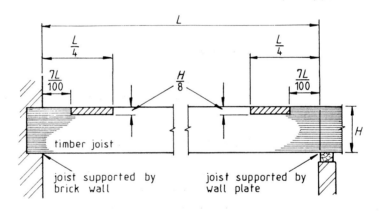

Figure 11.63   Dimensions for notches and holes

(a) Notches

Note Notches must be restricted to the hatched areas shown if the joists are not to be weakened.

L is the length of the structural member
H is the height of the structural member
If H exceeds 250 mm then for calculation purposes it is deemed to be 250 mm

Example For a joist 6 m long and 300 mm deep, notches must be:

(a) not more than $H/8$ deep, i.e.:

$$\frac{250}{8} = 31\,mm$$

(b) (i) at least $7L/100$ from bearing, i.e.:

$$\frac{7 \times 6000}{100} = \frac{42000}{100} = 420\,mm$$

(ii) not more than $L/4$ from bearing, i.e.:

$$\frac{6000}{4} = 1500\,mm = 1.5\,m$$

continued

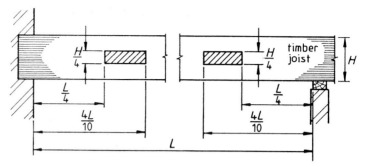

(b) Holes

Note Holes must be restricted to the hatched areas shown if the joists are not to be weakened.

L is the length of the structural member

H is the height of the structural member

If H exceeds 250 mm then use $H = 250$ mm for calculation purposes

Example For a joist 4 m long and 200 mm deep, holes must be:

(a) not more than $H/4$ in diameter, i.e.

$$\frac{200}{4} = 50 \text{ mm}$$

(b) (i) at least $L/4$ from bearing, i.e.

$$\frac{4000}{4} = 1000 \text{ mm} = 1 \text{ m}$$

(ii) not more than $4L/10$ from bearing, i.e.

$$\frac{4 \times 4000}{10} = 1600 \text{ mm} = 1.6 \text{ m}$$

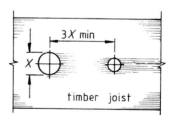

(c) Spacing of holes

The distance between adjacent holes in joists should be at least three times the diameter X of the larger hole.

Holes should pass through the centre of the joist and be parallel.

Figure 11.63 **Dimensions for notches and holes**
continued

Penetration of fire walls and floors

As required by the current Building Regulations, penetration of compartment walls and floors and fire barriers must be fire-stopped to prevent the passage of smoke and flame.

11.14 Jointing of pipework for potable water

Recent research has been carried out with the aim of improving water quality by reducing water contamination caused by bacteria or other harmful substances in mains or services.

This research, and the testing of water samples from mains and services, has shown that some traditional jointing materials harbour or promote the growth of bacteria. As a result, water authorities are now banning some of these materials and discouraging the use of others.

For example, bacterial contamination may be caused by linseed oil-based compounds commonly used with screwed joints. New compounds now available do not promote bacterial growth.

Again, if soldered joints are badly made, lead can be leached into solution and consumed. So, the water byelaws now permit only lead-free solder in capillary joints for hot and cold water supplies.

However, there are problems when choosing jointing materials because for some applications no suitable alternatives are available for the traditional materials.

Table 11.14 from BS 6700 lists permitted jointing materials and gives guidance on their use.

Table 11.14   Jointing of potable water pipework

Table 11.14 lists the jointing methods and materials that should be used for jointing potable water pipework. Materials and products that have been assessed by the UK Water Fittings Byelaws Scheme and listed by them are considered to meet the requirements of this table.

Type of joint	Method of connection	Jointing material	Precautions/limitations
Lead to brass, gunmetal, or copper, pipe or fitting	Plumber's wiped soldered joint	Tallow Flux	Of very limited application See note
Copper to copper pipe (pipe to pipe or fitting, or fitting to fitting)	Capillary-ring or end feed, soldered joint	Flux	No solder containing lead to be used See note
Copper to copper (pipe to fitting or fitting to fitting)	Non-manipulative compression fittings	—	Above ground only
	Manipulative compression fittings	Lubricant on pipe end when required	See note

continued

Table 11.14 **continued**

Type of joint	Method of connection	Jointing material	Precautions/limitations
Copper to copper (pipe to fitting or fitting to fitting)	Bronze welding or hard solder	Flux	See note
Galvanized steel (pipe to pipe or fitting), including copper alloy fittings	Screwed joint, where seal is made on the threads	PTFE tape or proprietary sealants	PTFE tape only up to 40 mm ($1\frac{1}{2}$) diameter See note
Galvanized or copper (pipe to pipe or fitting)	Flanges	Elastomeric joint rings complying with BS 2494, or corrugated metal. Vulcanized fibre rings complying with BS 216 or BS 5292	See note
Long screw connector	Screwed pipework with BS 2779 thread	Grummet made of linseed oil-based paste and hemp	See note
Shouldered screw connector	Seal made on shoulder with BS 2779 thread	Elastomeric joint rings complying with BS 2494 and plastics materials	—
Unplasticized PVC (pipe to fitting)	Solvent welded in sockets	Solvent cement complying with BS 4346: Part 3	—
	Spigot and socket with ring seal. Flanges. Union connectors	Elastomeric seal complying with BS 2494. Lubricants	Lubricant should be compatible with the unplasticized PVC and elastomeric seal
Cast iron (pipe to fitting)	Caulked lead	Sterilised gaskin yarn/ blue lead	See note
	Bolted or screwed gland joints	Elastomeric ring complying with BS 2494	—
	Spigot and socket with ring seal	Elastomeric seal and lubricant	—
Copper or plastic (pipe to tap or float-operated valve)	Union connector	Elastomeric or fibre washer	—
Stainless steel (pipe to pipe or fitting, including copper alloy fittings)	Soldered socket and spigot	Flux (phosphoric acid based) complying with BS 5245	No solder containing lead to be used See note

Table 11.14 continued

Type of joint	Method of connection	Jointing material	Precautions/limitations
Stainless steel (pipe to fitting, including copper alloy fittings)	Non-manipulative compression fittings	Lubricant when required	See note
	Manipulative fittings	Elastomeric seals when required	—
Pipework connections to storage cisterns (galvanized steel, reinforced plastics, polypropylene, polyethylene)	Tank connector/union with flange backnut	Washers: elastomeric, polyethylene, fibre	—
Polyethylene (pipe to fitting)	Non-manipulative fittings	—	Do not use lubricant
	Thermal fusion fittings	—	—
Polybutylene (pipe to pipe or fitting)	Non-manipulative fittings	Lubricant on pipe end when required	Lubricant if used should be listed and compatible with plastics
	Thermal fusion fittings	—	—
Polypropylene (pipe to pipe)	Non-manipulative fittings	Lubricant on pipe end when required	Lubricant if used should be listed and compatible with plastics
	Thermal fusion fittings	—	—
Cross-linked polyethylene (pipe to fitting)	Non-manipulative fittings	Lubricant on pipe end when required	Lubricant if used should be listed and compatible with plastics
Chlorinated PVC (pipe to fitting)	Solvent welded in sockets	Solvent cement	—

Note. Where non-listed materials are to be used, due to there being no alternative, the procedure used should be consistent with the manufacturer's instructions taking particular note of the following precautions:

(1) use least quantity of material to produce good quality joints;
(2) keep jointing materials clean and free from contamination;
(3) remove cutting oils and protective coatings, and clean surfaces;
(4) prevent entry of surplus materials to waterways;
(5) remove excess materials on completion of the joint.

11.15 Disinfection of an installation

Mains

Piping must be effectively cleansed and disinfected before being used and after being opened up for repair or alteration (see figure 11.64).

At the time of laying, large pipes should be brushed clean and sprayed internally with a strong solution of sodium hypochlorite. For small diameter pipes insert a polyurethane foam plug soaked in sodium hypochlorite solution at a strength of about 10% chlorine and pass it through the bore.

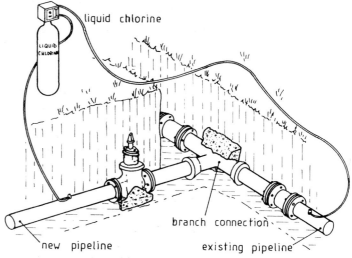

Before connection:
- ○ brush clean and disinfect pipes before insertion into pipeline;
- ○ disinfect trench around connection before cutting into existing pipeline;
- ○ put chlorine tablet or similar inside pipe before final connection.

Pipe from existing main can be used to fill new main for pressure test and for chlorination purposes, provided precautions are taken against backflow.

After connection:
- ○ flush out new main using water from existing pipeline;
- ○ fill with chlorinated water (50 ppm);
- ○ allow chlorinated main to stand for 24 hours;
- ○ flush out and take samples for chemical and bacteriological analysis.

Figure 11.64   Disinfection of mains

In pipework under pressure, chlorine should be pumped in through a properly installed injection point at the beginning of the pipeline, until the residual at the end of the pipeline is in excess of two parts per million.

If the pipework is under mains pressure inform the water supplier of the intention to carry out chlorination. All work must be carried out in accordance with the water supplier's requirements.

Installations within buildings

Storage cisterns and distributing pipes must be disinfected using the following method.

The cistern and pipe must first be filled with water and thoroughly flushed out. The cistern must then be filled with water again and a disinfecting chemical containing chlorine added gradually whilst the cistern is filling, to ensure thorough mixing. Sufficient chlorine should be used to give a concentration of at least 50 parts per million.

When the cistern is full, the supply should be stopped and all the taps on the distributing pipes must be opened successively, working progressively away from the cistern. Each tap should be closed when the water discharged begins to smell of chlorine. The cistern should then be topped up with water from the supply pipe and with more disinfecting chemical in the recommended proportions. The cistern and pipes should then remain charged for at least 3 hours, after which a test must be made for residual chlorine. If none is found, the disinfecting process must be repeated.

Finally, the cistern and pipes should remain charged with the chlorinated solution for at least 24 hours and then be thoroughly flushed out with clean water before any water is used for domestic purposes.

If ordinary 'bleaching powder' is used, the proportions must be 150 g of powder to 1000 l of water; the powder should be mixed with water in a separate clean vessel to a creamy consistency before being added to the water in the cistern. For proprietary brands of chemicals, use the proportions recommended by the manufacturer.

11.16 Testing

Smaller services should be tested by filling up the system at normal working pressure and inspecting all joints and fittings for leakage.

Any pipes below ground, buried under screeds, or in other inaccessible places should be tested before being covered. Hot water pipes should be similarly checked after heat has been applied. See chapter 12 for further information on testing.

11.17 Identification of valves and pipes above ground

Valves on hot and cold pipes should be fitted with an identification label of non-corrodible and non-combustible material, as shown in figure 11.65. The label should describe the function of the valve.

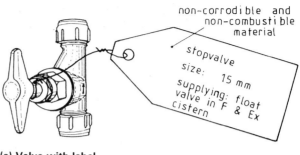

(a) Valve with label

Key

Control valve	Pipe run
SV1 SV2 SV3 SV4 SV5 }Servicing SV6 }valve SV7 SV8	sp1 sp2 }supply + sp3 }pipe sp4 dp1 }distributing dp2 }pipe cf1 }cold feed cf2 }pipe
Cisterns	
CWSC No. 1 }Combined feed and storage CWSC No. 2 }cistern (linked) F & Ex C Feed and expansion cistern	

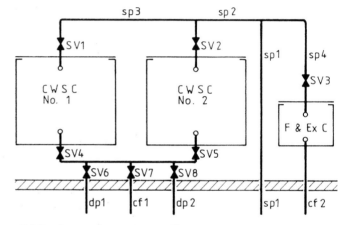

(b) Services and components diagram

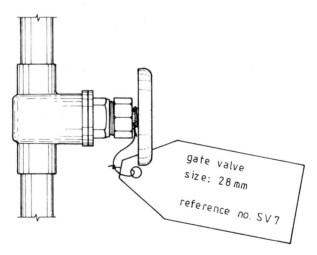

(c) Valve with label and identification reference number

Figure 11.65   Identification of pipes and services

Pipes within buildings should be colour banded as shown in figure 11.66, to identify the pipe and the service for which it is used.

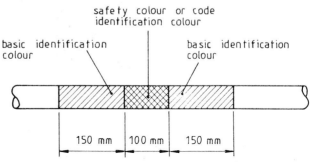

Colour code identification may be one or more colours.
Dimensions show minimum length for each colour.

(a) Colour coding

pipe contents	basic colour	colour code indication			basic colour
drinking water	green	auxiliary blue			green
boiler feed	green	crimson	white	crimson	green
central heating	green	blue	crimson	blue	green
cold down service	green	white	blue	white	green
hot water supply	green	white	crimson	white	green
fire extinguishing	green	red			green

(b) Example of colours used for water pipelines

Figure 11.66   **Identification of pipelines above ground to BS 1710**

Chapter 12
Inspection, testing and maintenance of pipelines, services and installations

Inspection and testing of pipes and installations is an integral part of the work and should be allowed for when estimating costs and during installation. It should be undertaken as the work proceeds.

Inspection and testing should ensure that:

(1) materials and equipment conform to British Standards or other forms of approval;
(2) installations are in accordance with the specification;
(3) byelaws and regulations are complied with.

It is important to notify water suppliers in time for them to witness any tests that they wish to see.

12.1 Timing of tests

Interim tests – as soon as practicable after completion of the particular section, with special attention to all work which will be concealed.

Final tests – to be carried out on completion of all work on water services, before handing over.

Items failing any test should be corrected immediately and retested before further work proceeds.

Visual inspection is an essential part of both interim and final tests and will detect many faults that the formal test will not pick up and which might lead to failure at a later date.

A careful record should be kept of such inspections and notes should be made to help with the preparation of 'as installed' drawings.

12.2 Inspection of below ground installations

During the visual inspection of pipelines, particular attention must be paid to the pipe bed, the line and level of the pipe, irregularities at joints, the correct fitting of air valves, washout valves, sluice valves and other valves together with any other mains equipment specified. This should include the correct installation of thrust blocks, where required, and ensure that protective coatings are undamaged.

Trenches must be inspected to ensure that excavation is to the correct depth to guard against frost and mechanical damage due to traffic, ploughing or other agricultural activities on open land.

Trenches should not be backfilled until these conditions have been satisfied and the installation is seen to conform to the drawings, specifications and appropriate byelaws and regulations.

12.3   Inspection of installations within buildings

All internal pipework must be inspected to ensure that it has been securely fixed.

Before testing takes place all cisterns, tanks, hot water cylinders and water heaters must be inspected to ensure that they are properly supported and secured, that they are clean and free from swarf, and that cisterns are provided with correctly fitting covers.

Byelaws note   Water undertakers expect to be notified before water installations are carried out, particularly installations below ground and those with a high contamination risk. Proper notification as in Byelaws 97 and 98 allows inspectors to discuss situations prior to installation and so avoid problems later. This also enables them to ensure that completed installations comply with relevant requirements.

12.4   Testing

Mains should be pressure tested at one and a half times to twice the normal working pressure (see figure 12.1). BS 6700 suggests twice working pressures but this may not always be advisable where working pressures are very high or for uPVC mains, which may retain stresses.

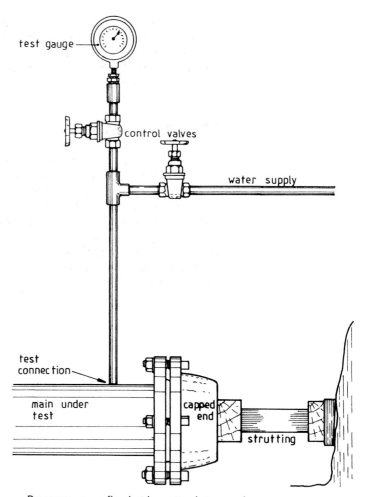

Pressure gauge fixed at lowest point on main.

Check and calibrate pressure gauge before beginning test.

Air valves positioned at high points to release trapped air.

Place sufficient backfill before testing to prevent movement of pipes under test.

Bring main up to test pressure slowly and release carefully after testing.

Using mains water fill main slowly to expel air and disconnect from main before beginning test.

Allow time for water absorption before starting test.

Capped ends strutted securely against solid ground.

Figure 12.1   Testing underground pipelines

Interim tests should be applied to *every* pipeline. For buried pipelines these must be carried out before backfilling is placed over the joints.

Final tests should be carried out only when all relevant work is complete. Completion of buried pipelines includes backfilling, compaction and surface finish.

12.5 Maintenance

To keep the performance of any installation at the original specified design standard, it will be necessary to carry out some maintenance work. The amount will depend upon the type and size of system, and the risk or effects of breakdown balanced against the frequency and cost of the inspection and maintenance programme.

Planned preventative maintenance, regularly carried out, will help to ensure that systems perform correctly and avoid most breakdowns and the risk of costly damage to components, equipment and buildings. Table 12.1 shows a typical maintenance schedule.

Table 12.1   Maintenance schedule

Component	Maximum time interval	Remarks
Inspections	12 months	At frequent intervals in addition to any statutory inspection.
Meters	6 months	Read meters for consumption and early warning of wastage. Check that meters are working.
Meter and stopvalve chambers	12 months or as required	Inspect to ascertain state of chamber construction. Clean out as necessary, check that box and lid are in working order and undamaged and grease hinges.
Water analysis	6 months	For drinking water storage of more than 1000 l.
Earthing and bonding		Check when other inspections, maintenance or alterations are being made (electrician to check electrical continuity).
Control valves	12 months	Operate to check that they close tightly and operate smoothly. Repair or renew as necessary. Valve key should be available for emergency use. Valves should be labelled clearly to show what they do.
Pressure relief valves and temperature relief valves	6 to 12 months	Ease at regular intervals to ensure they are not stuck or that outlet is blocked. Remedy faults immediately. Easing valves may sometimes cause them to leak but this is preferable to an explosion. Check discharge pipe is unobstructed.
Pressure reducing valves	6 to 12 months	Check pressures downstream of valve and investigate any changes from normal.
Vessels under pressure	6 months	Inspect water storage vessels and expansion vessels for signs of deterioration. Measure gas pressures and adjust if not within manufacturer's recommended limits.
Storage cisterns	6 months	Inspect for cleanliness and clean out as required. Check lid is securely in place. Look for signs of leakage or corrosion. Check linked cisterns for stagnant water (taste, odour or dust on surface). Check condition of bearers, safes and safe outlets. Check overflow pipes for obstruction and correct fall. Check insulation before winter.
Float-operated valves	6 months	Check operation and closing. Adjust for correct water level. Open float valve in feed and expansion cistern to prevent it sticking in the closed position.
Terminal fittings	6 to 12 months or as required	Rewasher, reseat or renew taps as required to prevent leakage. Tighten packing glands and check spindles for wear and efficient action. Check self-closing taps at intervals and remedy any faults. Clean sprayheads on taps, showers and shower mixers at intervals depending on rate of furring.

Table 12.1 continued

Component	Maximum time interval	Remarks
Pipework	12 months	Tighten loose fittings and supports and replace any missing ones. Check provision for expansion and contraction (especially plastics pipework). Check and tighten joints, remake or renew as necessary. Before commencing any work check compatibility of pipes and fittings, i.e.: ○ sizes, and whether imperial or metric; ○ metals for corrosion; ○ plastics for jointing method; ○ materials are available for repairs.
Corrosion	12 months	Inspect for outward signs of corrosion. Reduced flow rates may indicate that corrosion products are causing obstruction. Replace corroded or seriously scaled lead pipe with pipe of some other suitable material. Internal corrosion of galvanized steel pipe is usually localized, although in some waters the zinc coating can break up looking very like sand deposits in cisterns and tanks. In this case replace the complete pipe length. Pitting corrosion of copper pipe may be due to carbon film in pipe bore or cathodic scale. Rapid water velocities may lead to erosion corrosion. Pipes in damp conditions and areas where the air is acidic need regular inspections, and pipes should be protected against corrosive effects. Where pipes enter buildings they are particularly vulnerable at floor level, especially in the back of sink units.
Insulation and fire stopping	12 months	Inspect and make good any damage.
Ducting	6 months	Accessibility is essential. Check that ducting is clear of extraneous matter and free from vermin. Check that access is not obstructed and entry is readily possible. Check crawlways and subways for leakage from pipework, entry of ground or surface water, and accumulation of flammable materials.

12.6 Locating leaks

BS 6700 deals with leak detection to premises supplied by meter only. As most premises in this country are unmetered the author has also included notes relating to these.

The first signs of leakage in any premises are usually:

- Visual.
- Noise. The sound of water escaping through a small crack or hole in a pipe is similar to that of a cistern filling, but of course it will be continuous, and will probably be noticed more at night when the pressures are higher and there are less extraneous noises.
- In metered supplies the high meter reading or account is often the first indication, which leads to leak detection taking place.

Water Authorities periodically carry out their own waste water detection exercises and many leaks are located in this way. Water undertakers have a duty to keep wastage to a minimum, and if asked will usually send an inspector along to advise, and possibly to assist in locating both the service pipe and the leak. However, there are fairly simple ways to establish first that there is a leak, and secondly its approximate location.

Unmetered supplies

Procedure
(1) Look for visible signs of leakage, i.e.:
- wet or soggy patches on ground,
- areas of grass that are greener or growing stronger and quicker than remainder,
- water in stopvalve chambers.

(2) Make sure no water is being drawn off or used.

(3) Listen at taps and stopvalves to see where the noise appears loudest. There are excellent pocket stethoscopes available which can be used directly on a tap or on a stopvalve key which in turn is resting on an underground stopvalve. This will often give a general indication of the whereabouts of the leak.

(4) Turn off at the main stopvalve and listen again. If the noise has stopped, it usually means the leak is on the supply pipe. If the noise continues the leak is probably on the communication pipe or main.

Note It is important that the stopvalve turns off effectively, otherwise the leak may not be where it appears to be.

(5) If there are branch supplies with stopvalves fitted, then the procedure can be repeated by closing each branch supply in turn, to establish if the leak is on the common pipe or one of its branches, (see figure 12.2).

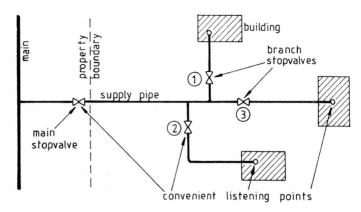

Figure 12.2   **Locating leaks in branch supply pipes**

(6) It is not always possible to locate the leak exactly, and in long pipelines the best approach may be to cut and plug the pipe about halfway along its length (or fit an intermediate stopvalve) and retest. This can be repeated a number of times to establish which length of pipe needs to be dug up, (see figure 12.3).

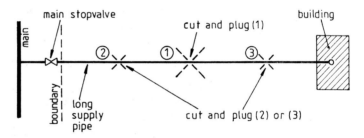

Figure 12.3   **Locating leaks in long pipelines**

(7) It is also possible to drive a pointed steel bar into the ground at intervals along a pipeline, pulling it out again to see how wet it is. Usually it is wetter the nearer the leak.

Note   The occurrence of a leak is often an indication of the state of the pipe, especially with steel pipelines.

Metered supplies

The method shown in BS 6700 follows a simple logical procedure using the meter and a watch, and requires first that the supply pipe and its stopvalves are located and recorded on a diagram. The isolating stopvalves should be numbered. The following example (see figure 12.4), based on that shown in BS 6700, describes the procedure.

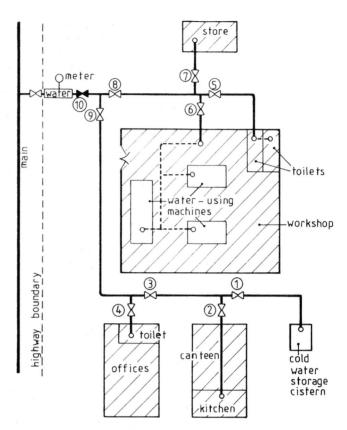

Figure 12.4   Locating leaks in metered supply – typical pipework network

Before the test:

- check that all stopvalves are in working order and will stop the supply when closed;
- make sure that no part of the supply is in use and that all water fittings except isolating valves are closed;
- check that the meter is capable of recording low flows.

Procedure

(1) Record rate of flow using the meter and a watch; if zero there is no detectable leakage.

(2) If there is a rate of flow, shut off isolating valves in sequence starting at the valve furthest from the meter, i.e. no. 1 in figure 12.4. After each valve has been shut any change in the rate of flow should be noted.

(3) Leakage should be sought in any sections of the network where closure of the isolating valve has reduced the rate of flow.

(4) As a check, repeat the test procedure when the leakages have been detected and repaired.

Practical example (see figure 12.4)

Item	Rate of flow l/min	Change in rate of flow l/min	Remarks
Start of test	90	—	
Shut valve no. 1	90	Nil	
Shut valve no. 2	45	45	Leakage at 45 l/min (in canteen section)
Shut valve no. 3	45	Nil	
Shut valve no. 4	45	Nil	
Shut valve no. 5	30	15	Leakage at 15 l/min (in toilets section)
Shut valve no. 6	30	Nil	
Shut valve no. 7	Zero	30	Leakage at 30 l/min (in supply to store)
Shut valve no. 8	Zero	Nil	
Shut valve no. 9	Zero	Nil	

12.7 Occupier information

Occupiers and owners of property have an interest in keeping their systems in good order and are in the position, as users, to note the first signs of faults appearing. Occupiers should be informed of the need for maintenance and be provided with maintenance instructions, and record drawings upon which pipe runs and valves are accurately marked.

British Standards referred to in text

BS 417	Galvanized low carbon steel cisterns, cistern lids, tanks and cylinders
	Part 2 Metric units
BS 486	Asbestos-cement pressure pipes and joints
BS 699	Copper direct cylinders for domestic purposes
BS 864	Capillary and compression tube fittings of copper and copper alloy
	Part 2 Capillary and compression fittings for copper tubes
BS 1010	Draw-off taps and stopvalves for water services (screw-down pattern)
BS 1211	Centrifugally cast (spun) iron pressure pipes for water, gas and sewage
BS 1212	Float-operated valves (excluding floats)
	Part 1 Piston type
	Part 2 Diaphragm type (brass body)
	Part 3 Diaphragm type (plastics body) for cold water services
BS 1387	Screwed and socketed steel tubes and tubulars and plain end steel tubes suitable for welding or for screwing to BS 21 pipe threads
BS 1394	Stationary circulation pumps for heating and hot water service systems
	Part 1 Safety requirements
	Part 2 Physical and performance requirements
BS 1566	Copper indirect cylinders for domestic purposes
	Part 1 Double feed indirect cylinders
	Part 2 Single feed indirect cylinders
BS 1710	Identification of pipelines and services
BS 1968	Floats for ballvalves (copper)
BS 1972	Polythene pipe (Type 32) for above ground use for cold water services
BS 2035	Cast iron flanged pipes and flanged fittings
BS 2456	Floats (plastics) for ballvalves for hot and cold water
BS 2494	Elastomeric joint rings for pipework and pipelines
BS 2580	Underground plug cocks for cold water services
BS 2779	Pipe threads for tubes and fittings where pressure-tight joints are not made on the threads (metric dimensions)
BS 2871	Copper and copper alloys. Tubes
	Part 1 Copper tubes for water, gas and sanitation
BS 2879	Draining taps (screw-down pattern)
BS 3251	Indicator plates for fire hydrants and emergency water supplies
BS 3284	Polythene pipe (Type 50) for cold water services
BS 3505	Unplasticized polyvinyl chloride (PVC-U) pressure pipes for cold potable water
BS 3506	Unplasticized PVC pipe for industrial uses
BS 3955	Electrical controls for household and similar general purposes
BS 4127	Light gauge stainless steel tubes
	Part 2 Metric units
BS 4213	Cold water storage and feed and expansion cisterns (polyolefin or olefin copolymer) and cistern lids
BS 4346	Joints and fittings for use with unplasticized PVC pressure pipes
	Part 3 Solvent cement
BS 4622	Grey iron pipes and fittings
BS 4772	Ductile iron pipes and fittings
BS 4991	Propylene copolymer pressure pipe
BS 5154	Copper alloy globe, globe stop and check, check and gate valves
BS 5163	Predominately key-operated cast iron gate valves for waterworks purposes
BS 5245	Phosphoric acid based flux for soft soldered joints in stainless steel

BS 5292	Jointing materials and compounds for installations using water, low-pressure steam or 1st, 2nd and 3rd family gases
BS 5306	Fire extinguishing installations and equipment on premises
BS 5422	Use of thermal insulating materials
BS 5433	Underground stopvalves for water services
BS 5728	Measurement of flow of cold potable water in closed conduits
BS 6076	Tubular polyethylene film for use as protective sleeving for buried iron pipes and fittings
BS 6281	Devices without moving parts for the prevention of contamination of water by backflow
	Part 2 Type B air gaps
	Part 3 Pipe interrupters of nominal size up to and including DN 42
BS 6282	Devices with moving parts for the prevention of contamination of water by backflow
	Part 1 Check valves of nominal size up to and including DN 54
	Part 3 In-line anti-vacuum valves of nominal size up to and including DN 42
BS 6283	Safety devices for use in hot water systems
	Part 2 Temperature relief valves for pressures up to and including 10 bar
	Part 4 Drop-tight pressure reducing valves of nominal size up to and including DN 54 for supply pressures up to and including 12 bar
BS 6465	Sanitary installations
	Part 1 Scale of provision, selection and installation of sanitary appliances
BS 6572	Blue polyethylene pipes up to nominal size 63 for below ground use for potable water
BS 6675	Servicing valves (copper alloy) for water services
BS 6730	Black polyethylene pipes up to nominal size 63 for above ground use for cold potable water

Bibliography

Backsiphonage Report, HMSO, (1973)

BS 1192 *Construction drawing practice*, Part 1, *General principles* (1984), Part 3, *Symbols and other graphic conventions*, (1987) British Standards Institution

Building Regulations, HMSO, (1985)

Design guide, Institute of Plumbing (out of print)

Guidance on the application and interpretation of the Model Water byelaws, Department of the Environment, (1986)

Model Water byelaws, HMSO, (1986)

New Water byelaws – course notes, Water Industry Training

Plumbing engineering services design guide, Institute of Plumbing

Principles of laying water mains, Water Authorities Association

Service cores in high flats – cold water services, Ministry of Housing and Local Government, (1965)

Water supply byelaws guide, Water Research Centre in association with Ellis Horwood Ltd, Chichester

Byelaws, published by each water undertaker

Index